SpringerBriefs in Computer Science

SpringerBriefs present concise summaries of cutting-edge research and practical applications across a wide spectrum of fields. Featuring compact volumes of 50 to 125 pages, the series covers a range of content from professional to academic. Typical topics might include:

- A timely report of state-of-the art analytical techniques
- A bridge between new research results, as published in journal articles, and a contextual literature review
- A snapshot of a hot or emerging topic
- An in-depth case study or clinical example
- A presentation of core concepts that students must understand in order to make independent contributions

Briefs allow authors to present their ideas and readers to absorb them with minimal time investment. Briefs will be published as part of Springer's eBook collection, with millions of users worldwide. In addition, Briefs will be available for individual print and electronic purchase. Briefs are characterized by fast, global electronic dissemination, standard publishing contracts, easy-to-use manuscript preparation and formatting guidelines, and expedited production schedules. We aim for publication 8–12 weeks after acceptance. Both solicited and unsolicited manuscripts are considered for publication in this series.

More information about this series at http://www.springer.com/series/10028

Iván Palomares Carrascosa

Large Group Decision Making

Creating Decision Support Approaches at Scale

 Springer

Iván Palomares Carrascosa
School of Computer Science (SCEEM)
University of Bristol
Bristol, UK

ISSN 2191-5768 ISSN 2191-5776 (electronic)
SpringerBriefs in Computer Science
ISBN 978-3-030-01026-3 ISBN 978-3-030-01027-0 (eBook)
https://doi.org/10.1007/978-3-030-01027-0

Library of Congress Control Number: 2018959758

This Springer imprint is published by the registered company Springer Nature Switzerland AG
The registered company address is: Gewerbestrasse 11, 6330 Cham, Switzerland

To my parents, Tomás and María,
and my sister Miriam.

Acknowledgements

The author would like to express his sincere thanks to the following colleagues and friends who made their contribution to some sections of this book: Jaime Solano Noriega (Universidad de Occidente, Mexico)—Chap. 2; Zhibin Wu (Sichuan University, China)—Chap. 4; Hengjie Zhang (Hohai University, China)—Chap. 4.

Contents

About the Author

Iván Palomares Carrascosa is a Lecturer in Data Science and Artificial Intelligence with the School of Computer Science, University of Bristol, UK. Since November 2018, he is also Fellow Member of the Alan Turing Institute (https://www.turing.ac.uk), the UK's leading Institute on Data Science and Artificial Intelligence, where he and his team members investigate personalization methods for healthy living and smart cities applications. Ivan received his two MSc degrees in Computer Science (with Faculty and Nationwide Distinctions) and Soft Computing and Intelligent Systems (Hons), from the University of Jaen, Spain, and the University of Granada, Spain, in 2009 and 2011, respectively. He received his PhD degree in Computer Science with Nationwide Distinctions from the University of Jaen, Spain, in 2014. He currently leads the Decision Support and Recommender Systems research group at the University of Bristol, where he supervises PhD candidates, postdoctoral and visiting researchers. His research interests include data-driven and intelligent approaches for recommender systems, personalization for leisure, tourism and healthy habits in smart cities, large group decision making and consensus, data fusion, opinion dynamics and human-machine decision support. His research results have been published in top journals and conference proceedings, including *IEEE Transactions on Fuzzy Systems, Applied Soft Computing, International Journal of Intelligent Systems, Information Fusion, Knowledge-Based Systems, Data*

and Knowledge Engineering, and Renewable and Sustainable Energy Reviews, amongst others. He serves as a reviewer in numerous top-tier international journals in related areas to Decision Support Systems.

More information about Ivan and his research group activities can be found at: http://dsrs.blogs.bristol.ac.uk.

List of Figures

List of Tables

The original version of this book was revised. A correction to this book is available at
https://doi.org/10.1007/978-3-030-01027-0_7

Chapter 1
Introduction

Abstract This chapter provides a concise introduction to the book, by explaining the motivation behind its elaboration and pointing out the need for a comprehensive text on the state, progress made and open questions revolving around large-group decision making research. Some notes are also provided for the potential readership of the book. The chapter finalizes with an outline of the book structure into chapters.

1.1 Motivation

Decision making processes take place just anywhere around us, with our own daily life situations being of course no exception. From the choice of the most suitable house to move in to the selection of the ideal candidate for a job position, or the adoption of the safest decision to evacuate a village after a natural disaster, there exist myriad situations where we encounter a number of options or decision alternatives, and we need to select the "best" one(s) or rank them from best to worst one according to our judgments and experience. When a single decision is to be made jointly by a group of people, we have a so-called *Group Decision Making* problem, in which usually each participant has their own individual opinions, concerns or interests towards the existing alternatives, but their opinions must be somehow combined into a representative opinion for the group that leads to the best (and ideally most accepted) solution by its members. Over the last decades, many researchers in the areas of group decision making and consensus building have widely investigated models, methods and decision support systems aimed at assisting decision groups in these situations, with numerous satisfactory results.

There is undeniably a wealth of published research, handbooks and monographs on different aspects of group decision making and consensus, most of which concentrate on small group decision problems. However, owing to the nowadays rise of technological paradigms capable of accommodating decision making across much larger groups of participants, much of the research efforts and scholars' attention have recently and increasingly shifted towards decision problems involving such large groups. Many of the classical approaches to support small group decisions are often limited and unsuitable to handle the various added difficulties stemming from

© The Author(s), under exclusive licence to Springer Nature Switzerland AG 2018
I. Palomares Carrascosa, *Large Group Decision Making*, SpringerBriefs
in Computer Science, https://doi.org/10.1007/978-3-030-01027-0_1

the participation of large decision groups. For this reason, a considerable number of research efforts have instead been devoted to defining models and methodologies specifically for supporting group decisions at large scale. The present book aims at providing the reader with a broad and complete yet concise vision of these works, outlining the main trends, models, methods and practical applications developed for supporting *Large Group Decision Making* processes to date, in an ample variety of real-life scenarios and domains.

In short, this book constitutes—to the best of the author's knowledge—the first central point of reference for the interested reader in the young but rapidly evolving field of large group decision making and consensus building in large groups. The book firstly provides an introduction to the broad research area of group decision making and consensus, after which the main characteristics and challenges of large group decision making (compared to conventional small-group decision making) are presented. Subsequently, the text focuses on providing a comprehensive literature review of related research in the topic, classified into six main trends. Related disciplines and notable application domains of large group decision making research are also highlighted throughout the book, along with final notes on proposed directions for future research.

1.2 Who Should Read This Book and Why?

The text provided in this book is primarily envisaged as a key point of reference for scholars, academics and research students across the communities of *Group and Multi-Criteria Decision Making under Uncertainty*, *Decision and Management Sciences*, and *Decision Support Systems*, along with scientists and practitioners from any of the numerous application areas of Group/Multi-Criteria Decision Aid approaches. For the acquainted reader with these areas, the book is aimed at providing a valuable reference point to a wealth of state-of-the-art work on large-group decision making, enabling a proper insight into:

- The existing literature with associated publications,
- active authors and research groups in the area(s),
- main trends and problems addressed,
- its most widely considered real-world applications, and
- potentially promising ideas for future investigation.

Research students and early career researchers may also benefit from having such a point of reference. For the unfamiliar but interested reader with the topic, we have considered the inclusion of a detailed introduction to *Group Decision Making*, *Consensus Building Approaches* and a brief summary of popular methods and principles for *Multi-Criteria Decision Making*. These preliminary contents are presented in the second chapter of the book, before moving into Large Group Decision Making. In addition, although not essential it would be highly recommended for the reader to have (or acquire) some basic knowledge about fuzzy set theory and its extensions, fuzzy preference modeling and aggregation/fusion of information in order to optimally understand the detailed discussions provided in the book. Bibliographical details of suggested readings on these topics can be found throughout the second chapter.

Since some of the newest approaches, trends and real-life applications covered in the following chapters involve the use of Data Science, Analytics, Operational Research and other Soft Computing and Intelligent Techniques, the book can be likewise of potential interest to a variety of scientists across these broader fields (e.g. computer science, operations research, management, social and political sciences, statistics, psychology), whose relationship with decision making research is becoming increasingly stronger and will be repeatedly pointed out throughout the text.

1.3 Chapter Overview

To accommodate both the familiar reader and a relatively new audience to the research problems being addressed, as discussed above, the structure of this book has been carefully planned and set out as follows:

- **Chapter 2: Group Decision Making and Consensual Processes**. This foundation-oriented chapter introduces the basic concepts, ideas and classical approaches proposed in the literature to support group and consensual decision making process. It also includes a basic overview of some important underlying steps to such processes, e.g. (1) the modeling of uncertain preferences, (2) the aggregation of individual preferences to yield representative preferential information at collective level, used for making group decisions, and (3) an overview of two popular methods for handling multi-criteria decision making problems, which often intersect with group and consensus decision making problems.
- **Chapter 3: Scaling Things Up: Large Group Decision Making (LGDM)**. The paradigm shift from classical small group decisions to large-scale decisions is reflected in this chapter. It enumerates the main difficulties and challenges exhibited by conventional approaches to effectively and efficiently managing large group decisions, identifies the research trends recently adopted to cope with such difficulties, and briefly describes the potential relationship between LGDM and other disciplines.
- **Chapter 4: LGDM Approaches and Models: A Literature Review**. Based on the research trends introduced in Chap. 3, this chapter provides the reader with a detailed survey of the extant LGDM models and methodologies in the literature, subdivided into six trends, identifying within each trend different themes and specific aspects investigated by researchers in the field.

- **Chapter 5: Implementations and Real-World Applications of LGDM Research**. This chapter overviews a number of model implementations into decision support systems for large groups, and enumerates the key real-life application areas where the existing literature has been applied.
- **Chapter 6: Conclusions and Future Directions of Research**. The book finalizes drawing some conclusions and pointing out several promising directions for research in this domain.

Chapter 2
Group Decision Making and Consensual Processes

Abstract This chapter introduces the basic concepts and ideas behind Group Decision Making (GDM) problems under uncertainty, highlighting its core underlying processes—aggregation of information and alternative(s) selection—and preference modeling approaches. Consensus building principles and its numerous related approaches to support accepted group decisions are then introduced in detail. Finally, given the frequent co-occurrence of decision scenarios involving both groups of participants and multiple evaluation criteria, the chapter concludes with an overview of classic Multi-Criteria Decision Making (MCDM) methods.

2.1 Decision Making Under Uncertainty

Decision making constitutes a core mankind activity in human beings' daily lives: we constantly face situations in which several (often mutually exclusive) alternatives exist, and we need to either (1) choose the best or most suitable alternative, or (2) establish a ranking of the alternatives from the best to the worst one. The study and application of decision making processes has historically taken place across a vast range of disciplines, including: business, management, economy and finance, engineering, planning, medicine and psychology, to name a few. As a consequence of this variety of application domains, myriad decision making models have been defined, thereby consolidating the establishment of Decision Theory as an "umbrella" and solidly justified area of research [16, 54, 62, 115].

In essence, a decision problem consists of four basic elements:

1. One or several *objectives* to solve.
2. A set of *alternatives*, each of which represents one of the possible decisions to be made for achieving the objective(s) pursued.

The original version of this chapter was revised. A correction to this chapter is available at https://doi.org/10.1007/978-3-030-01027-0_7

© The Author(s), under exclusive licence to Springer Nature Switzerland AG 2018
I. Palomares Carrascosa, *Large Group Decision Making*, SpringerBriefs in Computer Science, https://doi.org/10.1007/978-3-030-01027-0_2

3. A set of *factors* or states of nature, defining the context where the decision problem formulated takes place.
4. A set of *utility values*, each of which are associated to a specific alternative and state of nature.

Depending on the context where the decision problem occurs, a decision making process may take in one of the following environments or contexts.

- **Certainty environment**: The utility value of each alternative is accurately known; it is clearly and objectively established "how good" each alternative is.
- **Risk environment**: This situation arises when the knowledge about each alternative is modeled by a probability distribution.
- **Uncertainty environment**: In this situation, there exists some uncertainty of non-probabilistic nature associated to the utility values of the alternatives, hence such utility is characterized in an approximate fashion.

Classical approaches from Decision Theory provide suitable methods for solving decision problems defined in a certainty and risk environment. However, these methods are not adequate to deal with decision problems defined under uncertainty of a non-probabilistic nature, where the information about the problem is vague and imprecise [9]. These situations are also known as decision making problems in a fuzzy context, or "fuzzy decision making" problems. Zadeh's fuzzy set theory and fuzzy linguistic approach, along with their extensions, have proven to constitute an effective and very widely used tool to deal with uncertain information in myriad real-world decision problems [176–179].

Decision making problems under uncertainty can be categorized according to different points of view, e.g. based on the following two factors:

- **Number of participants**: When only one participant or *expert* takes part in the decision problem, we have an *Individual Decision Making problem*. On the contrary, when several experts take part in the decision problem together, we have a *Group Decision Making* problem [93].
- **Number of evaluation criteria**: Some problems require assessing each alternative "as a whole", i.e. based on only one attribute or evaluation criterion (*Single-Criterion* or Single-Attribute decision making), whereas other problems consider it necessary to assess alternatives in terms of multiple—sometimes conflicting— evaluation criteria (*Multi-Attribute* or *Multi-Criteria* decision making problems).

For the sake of clarity, we hereinafter adopt the following terminology and abbreviations. Acronyms GDM and MCDM are used to refer, respectively, to group decision making and multi-criteria decision making problems. Likewise, situations in which both multiple participants and multiple evaluation criteria coexist are referred to as *Multi-Criteria Group Decision Making* (MCGDM) problems hereinafter.

2.2 Group Decision Making (GDM) Problems

Making a collective decision—i.e. solving a GDM problem—implies the participation of several experts in a decision problem (each of whom have their own ideas, attitudes, motivations and knowledge), who attempt to achieve a common solution to the problem. Decisions in which several experts take part may often lead to better and less biased solutions than those made by a single expert. A GDM problem is formally characterized by the following elements:

- The existence of a decision problem to solve.
- A finite set X of $n \geq 2$ *alternatives* or possible solutions to the problem.

$$X = \{x_1, \ldots, x_n\}$$

- A group E of $m \geq 2$ individuals or *experts*, who express their opinions on the set of alternatives X and attempt to find a common collective solution to the problem at hand:

$$E = \{e_1, \ldots, e_m\}$$

Each expert $e_i \in E, i = 1, \ldots, m$, provides her/his opinions on the available alternatives in X by means of a preference structure P_i. Depending on the nature of the GDM problem and the level of expertise, knowledge or uncertainty exhibited by participants, different types of preference structures and formats. Section 2.3 overviews the most common preference structures and domains utilized in the GDM literature.

The solution for a GDM problem can be obtained by applying either a direct approach or an indirect approach (as illustrated in Fig. 2.1). In a *direct approach*, the solution is directly obtained from the individual preferences of experts without constructing a social (collective) opinion first [55]. By contrast, in the *indirect approach* a social opinion or *collective preference* is determined a priori from the aggregation of individual opinions and subsequently utilized to find the solution for the GDM problem. Regardless of the approach considered, the classical resolution process for GDM problems consists of two stages, as reflected in Fig. 2.2 [118]:

1. *Aggregation phase*: The individual preferential information from experts is combined by using an aggregation operator [8, 145].
2. *Exploitation phase*: It consists in identifying the best alternative(s) as the solution to the problem, or establishing a ranking of them from the most to the least preferred alternative by the group.

Furthermore, different situations can be encountered within the participatory context of a GDM problem. Examples of such situations include e.g. collaboration *vs* competitiveness among participants, compatible or incompatible proposals involving different environments (governments, companies, social platforms) etc. Accordingly, depending on the context and situation, the process

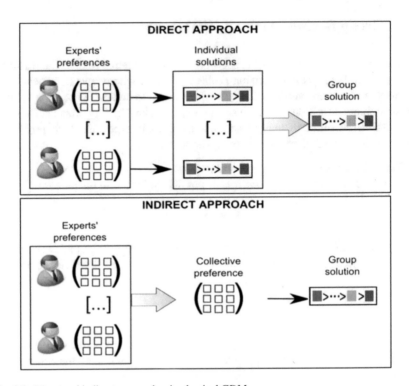

Fig. 2.1 Direct and indirect approaches in classical GDM

to find a solution for a GDM problem can be influenced by different guiding rules [18, 108]:

- *Majority rule*: The decision is made in accordance with the opinions of the majority of experts involved. Once the majority opinion has been adopted, it must be accepted and respected by other minority positions in the group, since it is assumed that all individuals accept a priori the use of this rule. The notion of majority admits two different modalities for its implementation:

  1. *Absolute majority*: The majority opinion has been adopted by more than half of the total number of experts in the decision group.
  2. *Relative or simple majority*: It only requires that the majority opinion is the one supported by the highest number of participants, even though the sum of the remaining experts supporting different opinions could be higher.

- *Minority rule*: The decision is delegated to a subgroup of individuals. This rule is frequently adopted for situations requiring a certain level of expertise that not all experts may have. It is therefore essential that all experts accept this rule and agree with the need for delegating the decision making process into a representative subgroup of them.

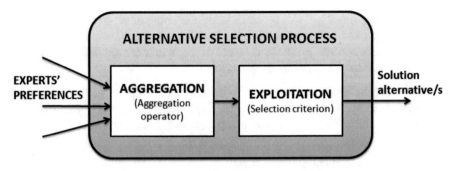

Fig. 2.2 Selection process for finding a solution in GDM problems

- *Individual*: This situation takes place when the group delegates the decision to an single person or there exists a leader in the group.
- *Unanimity*: All experts must agree with the decision made. Consensus-based approaches (introduced and revised in Sect. 2.4), are originally inspired by the concept of unanimity, although most of them consider a "softer" interpretation of unanimity, i.e. getting sufficiently close to unanimity without necessarily reaching it, as explained later.

2.3 Preference Modeling and Aggregation

As stated in the previous section, in a GDM each expert $e_i \in E$, $i = 1, \ldots, m$ provides her/his individual opinions about the existing alternatives in X by means of a preference structure. The most commonly utilized preference structures in related GDM (and MCGDM) literature are:

1. Preference orderings.
2. Utility vectors.
3. Preference relations.
4. Decision matrices.

They are formally defined below.

Definition 2.1 (Preference Relation [99]**)** A preference relation $P_i = (p_i^{lk})$ associated with an expert e_i on a set X of $n \geq 2$ alternatives, is defined by a fuzzy set on $X \times X$, represented by a $n \times n$ matrix of assessments $p_i^{lk} = \mu_{P_i}(x_l, x_k)$ as follows:

$$
P_i = \begin{pmatrix}
- & p_i^{12} & \cdots & p_i^{1n} \\
p_i^{21} & - & \cdots & p_i^{2n} \\
\vdots & \vdots & \ddots & \vdots \\
p_i^{n1} & \cdots & p_i^{n(n-1)} & -
\end{pmatrix}
\tag{2.1}
$$

with each assessment $p_i^{lk} \in D$ representing the strength of preference towards
alternative x_l when compared against another alternative x_k ($l \neq k$), according to
e_i. Clearly, assessments p_i^{ll} in the diagonal of the matrix are not defined, since an
alternative x_l cannot be compared against itself.

In the above definition, D denotes the information domain or preference format
utilized by experts to supply their opinions. Depending on the level of expertise and
uncertainty exhibited by the decision group members, D can be:

(a) *Quantitative*: numerical, e.g. a value in a discrete or continuous numerical scale
    or an interval-valued assessment, or,
(b) *Qualitative*: e.g. a linguistic assessment (as discussed later in this section).

For quantitative domains, two common types of preference relations are the additive
preference relation and the multiplicative preference relation. In the following
definitions, we introduce both types of preference relation as defined by Dong and
Xu's consensus building monograph [35].

Definition 2.2 (Additive Preference Relation [35]) An additive preference relation (also called fuzzy preference relation [99] or reciprocal preference relation) is
given by a matrix $P_i = (p_i^{lk})_{n \times n}$ such that $p_i^{lk} \in [0, 1]$, $l \neq k$, and $p_i^{lk} + p_i^{kl} = 1$.
If $p_i^{lk} > 0.5$, then x_l is strictly preferred against x_k. Conversely, if $p_i^{lk} < 0.5$
then x_l is strictly less preferred than x_k. Finally, if $p_i^{lk} = p_i^{kl} = 0.5$, then both
alternatives are equally preferred, i.e. the assessment indicates indifference between
x_l and x_k. Because of the additive reciprocity property, $p_i^{lk} > 0.5$ intuitively implies
$p_i^{kl} < 0.5$, and vice versa.

Definition 2.3 (Multiplicative Preference Relation [35]) A multiplicative preference relation is given by a matrix $P_i = (p_i^{lk})_{n \times n}$ such that $p_i^{lk} > 0$, $l \neq k$,
and $p_i^{lk} \cdot p_i^{kl} = 1$, with $\cdot$ denoting the usual product operator. Each assessment
indicates the ratio of preference intensity of alternative x_l with respect to x_l: if
$p_i^{lk} > 1$, then x_l is strictly preferred against x_k. On the contrary, if $p_i^{lk} < 1$
then x_l is strictly less preferred than x_k. Finally, if $p_i^{lk} = p_i^{kl} = 1$, then there
is indifference between x_l and x_k. Saaty's multiplicative scale[1] and its inverse
values are usually adopted to express assessments in multiplicative preference
relations.

Definition 2.4 (Preference Ordering [137]) A preference ordering associated
with an expert e_i is defined by the vector $O_i = (o_i^1, o_i^2, \ldots, o_i^n)^T$, with $o_i^l \in
\{1, \ldots, n\}$ the positional ranking index of alternative x_l in $X = \{x_1, \ldots, x_n\}$. The
closer this index to one, the better x_l is deemed against the rest of alternatives.

Definition 2.5 (Utility Vector) A utility preference (also referred to as utility
vector in the literature) associated with an expert e_i, is defined by the vector

---

[1]In Saaty's multiplicative scale, a value of 1 indicates indifference and the closer the integer value
is to 9 the more strongly x_l is preferred against x_k, see Table 2.1 in Sect. 2.5.

$U_i = (u_i^1, u_i^2, \ldots, u_i^n)^T$, where $u_i^l \in D$ is a value representing the assessment of the utility given by e_i to the lth alternative in X. The larger the value of u_i^l, the stronger the preference towards x_l.

Definition 2.6 (Decision Matrix) Given a finite set $C = \{c_1, \ldots, c_z\}$ of $z \geq 2$ evaluation criteria for the alternatives in X, a multi-criteria decision matrix—or simply decision matrix—associated with e_i on $X \times C$, is defined by a $n \times z$ matrix of assessments $M_i = (m_i^{lq})_{n \times z}$:

$$M_i = \begin{pmatrix} m_i^{11} & m_i^{12} & \cdots & m_i^{1z} \\ m_i^{21} & \cdots & \cdots & m_i^{2z} \\ \vdots & \vdots & \ddots & \vdots \\ m_i^{n1} & \cdots & \cdots & m_i^{nz} \end{pmatrix} \quad (2.2)$$

with $m_i^{lq} \in D$ representing the strength of preference towards x_l based on criterion c_q, according to e_i.

Quantitative (numerical) preference formats have been widely considered in many decision making approaches under uncertainty. Nevertheless, in many-real life decision making activities, it is not unusual that some elements need to be assessed qualitatively, because the experts may exhibit some degree of vagueness or imprecision, making it difficult for them to evaluate the available alternatives accurately. In other words, they would feel more comfortable by providing qualitative opinions, i.e. expressing their preferential information linguistically. The fuzzy linguistic approach and the concept of linguistic variable, introduced by Zadeh in [177–179], along with their numerous extensions to date [4, 97, 144] provide a solution to express opinions linguistically in a variety of fuzzy decision contexts [103]. We refer the interested reader to [95] for an in-depth study of the fuzzy linguistic approach and its most notable extensions, such as the popular linguistic 2-tuple model. A cornerstone element in any linguistic decision making process is the definition of a suitable linguistic term set with its associated syntax and semantics, describing the range of possible qualitative values (linguistic terms) that experts can utilize to express their preferences. A commonly adopted approach to do this is explained as follows. Let $S = \{s_h | h = 0, 1, \ldots, g\}$ be a finite linguistic term set with odd granularity[2] $g = |S|$, with $|S|$ the cardinality of the linguistic term set and $s_h \in S$ a linguistic term with its associated linguistic label. Then, S satisfies the following properties:

1. The set is ordered: $s_h > s_{h'}$, if $h > h'$.
2. There exists a negation operator: $s_h = neg(h')$ that satisfies $h' = g - h$.

---

[2]Existing linguistic decision making approaches tend to adopt two opposite notions of granularity: (1) odd granularity equal to cardinality of the term set, $g = |S|$; or (2) even granularity equal to $g = |S| - 1$.

Fig. 2.3 Linguistic term set with $g = 5$ linguistic terms

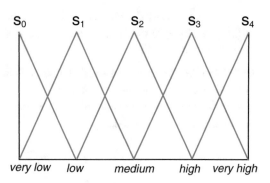

3. There is a maximum operator and a minimum operator: $\max(s_h, s_{h'}) = s_h$ if $s_h \geq s_{h'}$, and $\min(s_h, s_{h'}) = s_h$ if $s_h \leq s_{h'}$.

Note that according to the above concept of linguistic term set, its linguistic terms are symmetrically distributed on a scale under a total order relation. Furthermore, $s_{g/2}$ denotes the linguistic term situated in the middle of the ordered scale in S, interpreted similarly to a fuzzy assessment of "approximately 0.5" (or as indifference between alternatives in the case of assessments in a linguistic preference relation).

An example of linguistic term set with granularity $g = 5$ and the following linguistic labels, $S = \{s_0 = very\_low(VL),\ s_1 = low(L),\ s_2 = medium(M),\ s_3 = high(H),\ s_4 = very\_high(VH)\}$, is shown in Fig. 2.3. The semantics of the linguistic terms are given by fuzzy sets with triangular membership function (refer to [176] for more detail on the mathematical definition of fuzzy membership functions to characterize a fuzzy set).

Preference Aggregation The fusion of information is a fundamental process in virtually any decision aid models and decision support systems, thereby playing a key role in GDM and MCGDM processes [41]. The purpose of *aggregation functions*, also often called aggregation operators, is to combine a n-tuple of values or elements into a single representative value belonging to a domain (e.g. the unit interval [8, 170, 171] or other quantitative and linguistic preference domains including the ones outlined above). In GDM, the most obvious example of aggregation process pertains the fusion of individual preferences into a collective preference (previously shown in Fig. 2.2). Notwithstanding, it is also frequent in MCGDM models that, once individual preference matrices have been combined into a collective decision matrix, the elements in each row (assessments on a given alternative under several criteria) are subsequently aggregated to obtain a single representative assessment for each alternative.

Definition 2.7 ([8]) An aggregation function in the unit interval is a mapping $f : [0, 1]^n \to [0, 1]$, $n \geq 1$, producing an output value from a set of n input values $A = a_1, \ldots, a_n$. Every aggregation function in the [0,1] interval satisfies the following three properties:

1. **Identity when Unary**: $f(a) = a$.

Fig. 2.4 Summary and simplified spectrum of aggregation functions in the unit interval

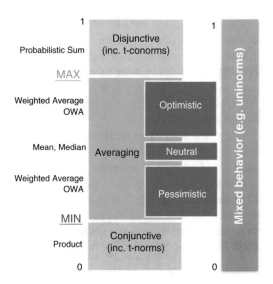

2. **Boundary**: $f(0, \ldots, 0) = 0$ and $f(1, \ldots, 1) = 1$.
3. **Monotonicity or Non-decreasing**: $a_z \leq b_z \ \forall z = 1, \ldots, n$ implies $f(a_1, \ldots, a_n) \leq f(b_1, \ldots, b_n)$.

Typically, the aggregation of preferential information in GDM and MCGDM models is conducted via averaging aggregation functions, in which $\min(a_1, \ldots, a_n) \leq f(a_1, \ldots, a_n) \leq \max(a_1, \ldots, a_n)$. However, as depicted in Fig. 2.4, the spectrum of aggregation functions also comprises operators with conjunctive or disjunctive behavior, i.e. operators in which the aggregated output is smaller than the minimum of the aggregation inputs (*conjunctive aggregation functions*) or the output is larger than the maximum of the inputs (*disjunctive aggregation functions*), respectively. Additionally, special types of aggregation functions such as the uninorm and nullnorm families of operators [171] are known as *mixed aggregation functions*, owing to their varying behavior which depends on the actual values of the inputs being aggregated.

Besides the above introduced basic properties of aggregation functions, in some specific decision contexts it is also desirable to utilize a function that fulfills additional mathematical properties, for instance:

1. **Idempotence**: $f(a, a, \ldots, a) = a$.
2. **Compensation**: $\min_z a_z \leq f(a_1, \ldots, a_n) \leq \max_z a_z$.
3. **Associativity**: $f(a, f(b, c)) = f(f(a, b), c)$.
4. **Reinforcement**: Tendency of multiple high (*resp.* low) values to reinforce each other, leading to an even higher (*resp.* lower) result.

For the interested reader, we refer to [8] for a comprehensive overview of the main classes of aggregation functions, and to [101, 109, 170, 171] for detailed discussions on some well-known aggregation operators in GDM. Although some basic examples of aggregation operators in the unit interval are described below, a great deal of research has been devoted (and continues being devoted [145]) to defining extensions of these into various preference formats, for instance linguistic assessments [55], intuitionistic fuzzy assessments [81], hesitant fuzzy assessments [157], etc. The elaboration on these domain-specific extension remains outside of the main scope of this text.

The arithmetic mean is arguably the simplest method to aggregate individual assessments into a collective assessment, in situations where all individuals opinions are deemed as equally important. For notation convenience, let us assume in the remainder of this section that each aggregation input a_i, $i = 1, \ldots, n$ represents an assessment (e.g. a preference value on a given alternative, pair of alternatives or alternative-criterion pair) provided by the ith expert in a group of size n. Then, an aggregated or collective assessment can be obtained by applying the arithmetic mean operator as follows:

$$a = AM(a_1, \ldots, a_n) = \frac{1}{n} \sum_{i=1}^{n} a_i \qquad (2.3)$$

The geometric mean is frequently used in conjunction with preferences expressed under Saaty's multiplicative scale.

$$a = GM(a_1, \ldots, a_n) = \left(\prod_{i=1}^{n} a_i \right)^{\frac{1}{n}} \qquad (2.4)$$

However, neither the simple (non-weighted) arithmetic mean nor the simple geometric mean operators would provide meaningful results if all or most experts exhibit extreme preferences in the assessment, in which case the aggregated output may not be representative of none of the individual opinions.

The weighted arithmetic mean and weighted geometric mean are a suitable approach to combine preferential information when the individuals (or criteria) have distinct importance weights, hence a weighting vector $W = [w_1 \ w_2 \ \ldots \ w_n]^T$ $w_i \in [0, 1]$ associated to the aggregation inputs, is introduced. In many decision situations such weights are required to be normalized, i.e. besides $w_i \in [0, 1]$, the weighting vector W also holds $\sum_i w_i = 1$. Let us assume in the sequel, without loss of generality, that weights are normalized.
Weighted arithmetic mean:

$$a = WAM_W(a_1, \ldots, a_n) = \frac{1}{n} \sum_{i=1}^{n} a_i \cdot w_i \qquad (2.5)$$

Weighted geometric mean:

$$a = WGM_W(a_1, \ldots, a_n) = \prod_{i=1}^{n} a_i^{w_i} \qquad (2.6)$$

The OWA (Ordered Weighted Averaging) family of operators were introduced by Yager in [170], and they constitute a widely used assortment of weighted aggregation operators in the fuzzy multi-criteria decision making and GDM literature [118], with interesting properties and a remarkable number of extensions being developed in the last decades, e.g. [55, 101, 172]. They are formally defined as follows:

Definition 2.8 ([170]) Let $A = \{a_1, \ldots, a_n\}$ $(a_z \in [0, 1])$ be a set of n aggregation inputs. A OWA operator is a mapping $OWA_W : [0, 1]^n \rightarrow [0, 1]$, with an associated weighting vector $W = [w_1 w_2 \ldots w_n]^\top$, such that $w_z \in [0, 1]$, $\sum_z w_z = 1$ and,

$$OWA_W(a_1, \ldots, a_n) = \sum_{z=1}^{n} w_z b_z \qquad (2.7)$$

where b_z is the z-th largest value in A. OWA operators are characterized by assigning a weight w_z to the z-th largest element in A, unlike classic weighted average operators, which assign a weight w_i to the i-th aggregation input, a_i.

The behavior of OWA operators (either optimistic, pessimistic or neutral, see Fig. 2.4) can be flexibly defined and classified based on their weighting vector W. To determine the attitudinal character of the specific operator being used, a measure called degree of optimism or *orness* was also introduced in [170]:

$$orness(W) = \frac{1}{n-1} \sum_{z=1}^{n} (n - z) w_z \qquad (2.8)$$

Optimistic (OR-like) OWA operators are those where $orness(W) > 0.5$, whereas pessimistic (AND-like) operators accomplish $orness(W) < 0.5$. The higher $orness(W)$, the more importance is assigned to the highest values in A, therefore the closer the output is to $max(a_1, \ldots, a_n)$. Conversely, the lower $orness(W)$, the more importance is given to the highest inputs in A, and the closer the output is to $min(a_1, \ldots, a_n)$.

A central aspect for the definition of an OWA operator consists in the construction of the weighting vector W. Different approaches have been proposed in the literature to facilitate their computation, e.g. by using fuzzy linguistic quantifiers or via learning approaches [101, 170, 171]. Some particular cases of OWA operators are:

- The *maximum* operator, with $orness(W) = 1$, $w_1 = 1$ and $w_z = 0$, $z \neq 1$.
- The *minimum* operator, with $orness(W) = 0$, $w_n = 1$ and $w_z = 0$, $z \neq n$.
- The *arithmetic mean*, with $orness(W) = 0.5$ and $w_z = 1/n$ $\forall z$.

Besides the aggregation of preferential information across individuals or criteria, some approaches [20, 109, 195] involve other distinct uses of aggregation functions requiring different behaviors from the averaging behavior. For instance, when attempting to capture the trend exhibited by an expert preference on an alternative across multiple time instants [20], or trying to reflect the evolution of the (measured) behavior of participants in a consensus building process [109] (introduced in the next section), we would be instead interested in *reinforcing* the presence of multiple high (resp. multiple low) aggregation inputs together. Uninorm aggregation operators are a clear example of mixed behaviour functions accomplishing this requirement. They were introduced by Yager and Rybalov in [171] to provide a generalization of the t-norm and the t-conorm operators [8]. Unlike t-norms and t-conorms, whose neutral elements are 1 and 0 respectively, uninorms have a neutral element $g \in [0, 1]$ lying anywhere in the unit interval. Whilst OWA operators allowed to define varying attitudes within an averaging behavior, uninorm aggregation operators present a varying behavior (namely conjunctive, disjunctive or averaging), depending on the input values being higher or lower than g. They are defined as follows:

Definition 2.9 ([171]) A uninorm is a mapping, $\mathscr{U} : [0, 1]^2 \to [0, 1]$, having the following properties for all $a, b, c, d \in [0, 1]$:

  i) *Commutativity*: $\mathscr{U}(a, b) = \mathscr{U}(b, a)$.
  ii) *Monotonicity*: $\mathscr{U}(a, b) \geq \mathscr{U}(c, d)$ if $a \geq c$ and $b \geq d$.
  iii) *Associativity*: $\mathscr{U}(a, \mathscr{U}(b, c)) = \mathscr{U}(\mathscr{U}(a, b), c)$.
  iv) *Neutral element*: $\exists g \in [0, 1] : \mathscr{U}(a, g) = a$.

Due to the associativity property, uninorm operators are typically defined for $n = 2$, and additional input values can be successively aggregated without affecting the aggregated result. The conjunctive, disjunctive or averaging behavior depends on input values a, b being (1) both lower than g, (2) both greater than g, or (3) one above and one below g, respectively. In practice, this translates into a notable characteristic of uninorm operators: their *full reinforcement* property. Given any $g \in [0, 1]$, uninorms show an *upward reinforcement* when both input values are high (above g), making the aggregated value even higher (disjunctive behavior). Conversely, they show a *downward reinforcement* when aggregating low input values (below g), so that the aggregated value is even lower (conjunctive behavior).

    An example of uninorm function is the cross-ratio operator, which is a continuous uninorm in $[0, 1]^2 \setminus \{(0, 1), (1, 0)\}$, with neutral element $g = 0.5$:

$$\mathscr{U}(a, b) = \begin{cases} 0 & \text{if } (a, b) \in \{(0, 1), (1, 0)\}, \\ \dfrac{ab}{ab + (1-a)(1-b)} & \text{otherwise.} \end{cases} \qquad (2.9)$$

2.4   Consensus Building in GDM

In most aforesaid GDM situations in which an alternative selection process is applied solely (aggregation and exploitation), it might happen that some experts do not accept the decision made, because they consider that their preferences have not been sufficiently taken into account. Ensuring that a high level of collective agreement is achieved, is an aspect of vital importance in many real-life GDM problems. Therefore, it becomes necessary to apply a Consensus Reaching Process (CRP), which introduces an additional phase prior to the selection process previously described in Fig. 2.2, with the aim of achieving a high level of collective agreement before making a group decision [18, 126].

The term *consensus* can be defined[3] as "a generally accepted opinion or decision among a group of people". In [126], Saint et al. defined consensus as "a state of mutual agreement between members of a group, where all legitimate concerns of individuals have been addressed to the satisfaction of the group". Most definitions for consensus assume the idea of a collective decision making process after which no experts disagree with the decision made, although some of them may still consider that their preferred solution would have been better than the actual solution found. In order to achieve consensus, it is often necessary that most or all experts modify their initial opinions, bringing them closer to each other, towards a collective opinion viewed as satisfactory by the group.

The concept of consensus has classically caused some controversy within the GDM community, since it can be subject to multiple interpretations, from a classical view of consensus as total agreement (*unanimity*) to more flexible interpretations. Consensus as unanimity [72] is in most practical situations difficult or even impossible to achieve. Likewise, such "unanimity" might have been achieved by means of intimidation or other external circumstances imposed on the group, so that no true agreement is really made: this type of consensus situation is referred to as *normative consensus* [143]. Rather than unanimous agreement or normative consensus, the notion of consensus should be rather understood as the result of an iterative and participatory discussion process in which the final decision made may not be in total accordance with the initial positions of the individuals. This view of consensus is known as *cognitive consensus*, and it implies that the experts modify their initial opinions after several rounds of discussion and negotiation [94]. Based on the concept of cognitive consensus, a number of flexible approaches for consensus building that consider different degrees of partial agreement, have been proposed in the literature [18, 56, 69]. One of the most accepted approaches to soften the conventional, yet strict view of consensus as unanimity, is the one called *soft consensus*, introduced by Kacprzyk in [67]. Based on the concept of fuzzy linguistic majority, this approach assumes the existence of consensus in a decision group when "most of the important

---

[3]Cambridge English Dictionary.

individuals agree as to (their testimonies concerning) almost all of the relevant options" [68, 69]. The concepts of *soft consensus* and fuzzy majority are based on fuzzy set theory [176] and fuzzy linguistic quantifiers [180], respectively, and their associated approaches have provided satisfactory results in numerous GDM frameworks [69–71].

In a CRP, the primary goal is to obtain a desired level of agreement before applying the alternatives selection process, after one or several rounds of discussion on preferences [126]. The CRP is an iterative and dynamic process usually coordinated by a human entity called *moderator*. The moderator is a key figure in such processes, whose main functions are [103]:

- To evaluate the level of existing agreement at each discussion round of the CRP.
- To identify the alternatives whose discrepant opinions among participants hamper achieving consensus.
- To inform participants about the changes they should consider on their preferences, regarding the alternatives identified.

Three basic assumptions shall be understood and accepted a priori by all participating experts before initiating a CRP:

- Every member of the group *must* understand the process carried out to achieve consensus, clarifying any possible questions or doubts before initiating their participation in it.
- Undertaking a CRP implies that all experts *accept* to collaborate with each other so as to find a collectively agreed solution.
- If required, experts *should move* from their initial positions, in order to bring their preferences closer to the rest of the group.

Figure 2.5 shows a general scheme followed by most existing approaches for consensus building in the literature. Its main stages are described below:

1. *Consensus measurement*: the individual preferences of all experts over the existing alternatives, P_i, $i \in \{1, \ldots, m\}$, are gathered to determine the current level of agreement in the group predicated on a consensus measure. The aim of

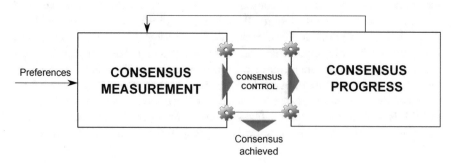

Fig. 2.5 General consensus building scheme

a consensus measure is therefore to quantify how close the opinions of experts are from unanimous agreement. Further detail about the broad types of existing consensus measures is provided later in this section.

2. *Consensus control*: the consensus degree computed in the previous phase is analyzed to decide whether it is sufficient or not to make a group decision against the problem at hand. If the consensus degree is enough, then the group moves on to the selection process. Otherwise, it is necessary to carry out another round of discussion or perform further actions to improve the consensus level. Two parameters, whose values are fixed a priori by the group, are often utilized in this phase:

   - A consensus threshold μ, whose value indicates the minimum level of agreement required amongst members in the group. Many consensus models compute the consensus degree as a value in the unit interval [59, 70, 114], with a value of 1 being interpreted as full agreement (unanimity), therefore $\mu \in [0, 1]$ in such cases. Intuitively, the higher μ, the more strict the need for a highly accepted decision across the entire group.
   - A maximum number of discussion rounds allowed, $Maxround \in \mathbb{N}$. If the number of rounds carried out exceeds this value, then the CRP ends without having reached consensus.

3. *Consensus Building*: If the current degree of consensus is not enough, a procedure is applied to increase the level of agreement in the following round of the CRP. Traditionally, this procedure is based on providing experts with some feedback, advising them how to modify their original preferences. However, some approaches that conduct this process automatically have been also proposed, for instance to accommodate time-sensitive decision problems in which an accepted decision must be found as quickly as possible:

   (a) *Feedback Mechanism*: This is the usual process carried out in classical CRPs, in which human experts discuss about their preferences, guided by a moderator (or a computer-based consensus support system acting as such [71]). The moderator identifies the farthest experts' assessments from consensus in the current round. Subsequently, these identified experts are provided with some advices to modify the value of assessments previously identified, so as to bring them closer to the rest of the group and increase the consensus degree in the next CRP round. Numerous consensus models incorporate feedback mechanisms based on this process [15, 21, 59, 96]. Figure 2.6 illustrates a general scheme for CRPs with feedback generation. Consensus approaches based on a feedback mechanism have the advantage of keeping experts in control of their own preferences throughout the CRP, i.e. their sovereignty is preserved. However, they sometimes present the disadvantage of requiring a considerable effort and temporal cost invested in manually adjusting assessments based on several iterations of feedback. This situation may aggravate in problems where most participants refuse to modify their opinions in accordance with the feedback received or they simply ignore it.

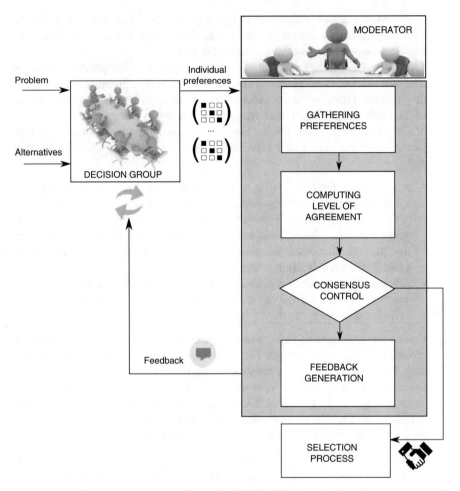

Fig. 2.6 General scheme of CRPs based on feedback mechanism

(b) *Automatic Adjustments*: Instead of incorporating a feedback mechanism, some consensus models implement approaches that update information (e.g. assessments of experts and/or their importance weights) to increase consensus in the group automatically [10, 35, 50, 150, 156, 160, 184]. Therefore, once experts provide their initial preferences at the beginning of the CRP, they do not need to manually supervise them at each round, neither by accepting/declining proposed changes in their opinions, nor manually updating them. These approaches are suitable for GDM problems whose main priority is to achieve a consensual decision upon minimum human effort, or in time-sensitive scenarios such as emergency decision making. On the contrary, they should not be adopted in problems where the sovereignty and "sense of control" assumed by participants on their own opinions must be preserved.

(c) *Hybrid or Semi-Supervised Approach*: In addition to the above, recent studies have defined semi-supervised or semi-automatic approaches that only require partial supervision of experts' assessments throughout the CRP, applying updates on such assessments automatically in certain circumstances, e.g. when the suggested advice on a given assessment does not involve an overall drastic change in the expert's opinion [104, 111].

The innumerable research efforts devoted to devising and improving consensus support approaches during the last few decades, have undeniably made it become a solid and well-established topic of research within GDM [35, 61, 103]. As a result, a large number of consensus approaches—both theoretical and practical—have been proposed by different authors, including:

1. *Consensus measures* [14, 27, 51, 57, 68, 69, 73], i.e. measures to compute the level of group agreement from the individual preferences of experts. The measurement of consensus has been investigated across diverse disciplines [51], and consequently different consensus measures have been formulated. Most consensus measures rely on applying similarity or distance metrics to compute the closeness between experts' preferences, as well as aggregation operators that obtain the global level of group agreement from the aggregation of similarity degrees previously calculated at assessment level [8].

2. *Consensus models* [15, 35, 56, 59, 60, 96, 105, 113, 114, 126, 160, 184] provide groups with the necessary guidelines and specifications to support them in undertaking a CRP in a variety of GDM frameworks. Numerous consensus models have been proposed by several researchers to support CRPs in different GDM contexts, including: (1) decision environments characterized by uncertainty and vagueness, that require the use of linguistic information domains suitable to express preferences [27, 56]; (2) MCGDM problems where alternatives must be assessed under several evaluation criteria [111, 113, 160]; (3) groups of experts with diverse background and expertise, who may require using different preference structures depending on their level of expertise [59]; or (4) scenarios characterized by social relationships between participants or a social network structure [88, 151], among others.

3. *Consensus Support Systems* (CSS) [21, 31, 71, 100], i.e. implementations of existing models into computer-based, Web-based or mobile-based decision support systems specifically aimed at supporting CRPs. Some of the benefits provided by CSS with respect to conventional CRPs are the partial or total automation of the tasks typically carried out by a human moderator, and the possibility of conducting non-physical meetings where participants may be geographically separated, with the aid of e.g. Web and mobile technologies.

2.4.1   Overview of Consensus Measures

Based on the literature review conducted in [103], it was reported that most consensus measures in the literature can be broadly classified into two types (see Fig. 2.7), depending on the nature of the computations and fusion procedures applied on individual preferential information.

1. *Consensus measures based on distances to the collective preference* [10, 56, 127]. The procedure to measure agreement in this type of consensus measures is as follows. Firstly, the collective preference, denoted by P_c and representing the global opinion of the group is computed by aggregating all individual preferences of experts, i.e. $P_c = \phi(P_1, P_2 \ldots, P_m)$, with ϕ an aggregation operator. The individual consensus degrees are then obtained by computing the distances between each individual preference and the collective preference, $d(P_i, P_c)$, with $d(\cdot, \cdot)$ a distance metric. The level of consensus in the group can subsequently be determined by aggregation of the individual consensus degrees.
2. *Consensus measures based on pairwise similarities (distances) between experts* [14, 57, 69, 70]: This type of consensus measure calculates the similarity between pairs of experts' opinions. For each pair of experts in the group, $(e_i, e_j), i < j$, the degrees of similarity between their opinions (assessments) are computed based on a distance metric. Similarity values $sim(P_i, P_j)$ are then aggregated to obtain consensus degrees at group level. Importantly, for a group with m members, this process implies calculating similarities between all the $m(m-1)/2$ different pairs of them.

One of the first precursors to "soft" consensus measures was proposed by Spillman et al. [127], based on mathematical procedures taken from fuzzy set theory [176]. In practice, this supposes an early effort to comply with a more realistic and flexible notion of consensus, as opposed to the classical view of consensus as unanimous agreement still being adopted a few years before, for instance in Kline's work [72]. Spillman et al. consider measuring the degree of consensus for each expert separately, as the distance between his/her pairwise comparisons among alternatives—given by a reciprocal fuzzy preference relation—and an "ideal" consensus matrix with maximum consensus degree, determined a priori predicated on matrix calculus. A second measure called fuzziness degree is also defined, whose value is larger if the consensus degree is lower and vice versa. Both the consensus and fuzziness degrees are utilized jointly to quantify the level of group agreement.

Later on, in the mid-90s Herrera et al. introduced an innovative consensus measure for linguistic preferences [56]. Their study constitutes one of the earliest efforts to cope with the situation in which experts might sometimes have a vague knowledge about the problem at hand, hence they would prefer to use linguistic assessments instead of numerical ones. Alternatives and experts have fuzzy importance degrees, inspired by Kacprzyk's soft consensus approach [69].

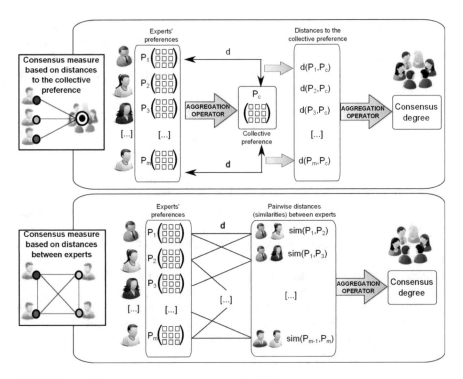

Fig. 2.7 Types of consensus measures as surveyed in [103]

Herrera et al.'s method calculates two types of consensus measure: *consensus degrees*, i.e. indicators of the current level of agreement; and *linguistic distances*, used to evaluate the distance from each expert's linguistic preference relation to the collective opinion in a linguistic manner, e.g. "very far from", "slightly near", etc. Both two measures are assessed linguistically by using linguistic terms s_h from a finite linguistic term set S defined a priori. Furthermore, the aforesaid consensus degrees and linguistic distances are calculated at three levels using the Linguistic OWA operator, an extension of classical unit interval-based OWA operators [55], to aggregate information expressed as linguistic term sets within a finite ordered set. This is done through the following three steps: (1) a counting process, (2) a coincidence process and (3) a computing process. Subsequently, in [58], the authors extended the linguistic consensus measures described above, by incorporating and applying a consistency control process on preferences. This process takes place before measuring consensus.

Ben-Arieh and Chen [11] investigated the problem of aggregating linguistic preferences expressed as fuzzy set membership functions in a common linguistic term set by a group of experts who have associated linguistic importance weights. They extended the Fuzzy-LOWA aggregation operator proposed in [10] to integrate the linguistic weights of individuals in the aggregation of individual preferences.

Accordingly, they defined a consensus measure in which individual preference orderings and a collective preference ordering (both derived upon their associated linguistic preference structures) are compared against each other. The following formula illustrates Ben-Arieh and Chen's process to calculate the consensus degree on an alternative x_l, denoted by ca^l:

$$ca^l = \sum_{i=1}^{m} \left[\left(1 - \frac{|O_i^l - O_c^l|}{n-1} \right) \times w_i \right] \qquad (2.10)$$

with O_i^l and O_c^l the ordered positions of alternative x_l in the preference orderings associated to e_i and the collective opinion, respectively, and w_i the importance weight of e_i. The arithmetic mean operator is then used to obtain the overall consensus degree by aggregating all $ca^l, l = 1, \ldots, n$.

Kacprzyk et al. proposed human-consistent measures of consensus that capture our perception of consensus in practice more faithfully than the strict notion consensus as unanimous agreement. As a result, they coined an alternative notion of soft consensus, based on the concept of fuzzy majority [67–69]. They introduced a consensus measure based on pairwise similarities between experts' additive preference relations, whereby the level of agreement is hierarchically computed at multiple levels, starting by α-degrees of sufficient agreement (with $\alpha \in [0, 1]$) on the assessments p_i^{lk} and p_j^{lk}:

$$sm_{ij}^{lk} = \begin{cases} 1 \text{ if } |p_i^{lk} - p_j^{lk}| \le 1 - \alpha \le 1, \\ 0 \text{ otherwise} \end{cases} \qquad (2.11)$$

2.4.2   Consensus Building Approaches

Once the degree of consensus has been determined, if it is insufficient then it becomes indispensable to implement some action to bring individuals' opinions closer to each other. As previously discussed, there are two main families of approaches for consensus building in the extant GDM literature: feedback mechanisms involving the active participation and engagement of experts throughout the CRP, and automatic approaches in which, once experts provide their initial opinions, consensus is built automatically without further human experts intervention. Taking this distinction into account, along with the previously introduced classification of consensus measures into two types, Palomares et al. [103] defined the taxonomy depicted in Figs. 2.8, 2.9 and 2.10 for categorizing the existing consensus research under contexts of fuzziness.

Feedback mechanisms for consensus building usually involve two major aims:

1. Identifying the experts whose opinions are farthest from the consensus (aggregated) opinion. This is frequently done by determining the proximity between

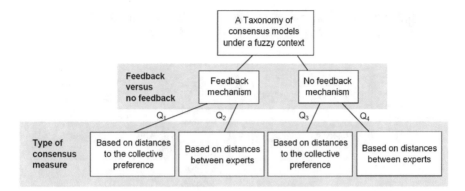

Fig. 2.8 Taxonomy of consensus research in a fuzzy context [103]

(Q1) FEEDBACK MECHANISM AND CONSENSUS MEASURES BASED ON DISTANCES TO THE COLLECTIVE PREFERENCE

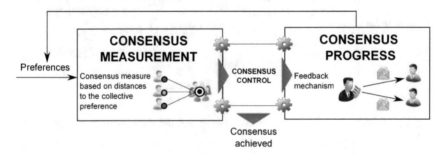

(Q2) FEEDBACK MECHANISM AND CONSENSUS MEASURES BASED ON DISTANCES BETWEEN EXPERTS

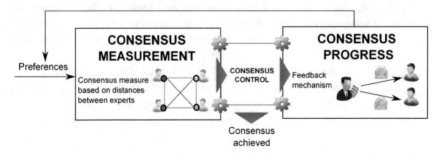

Fig. 2.9 Consensus models with feedback mechanism and different types of consensus measures

(Q3) AUTOMATIC ADJUSTMENTS AND CONSENSUS MEASURES BASED ON DISTANCES TO THE COLLECTIVE PREFERENCE

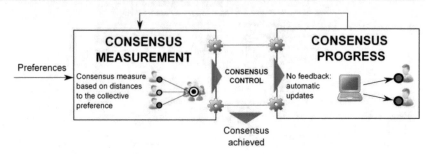

(Q4) AUTOMATIC ADJUSTMENTS AND CONSENSUS MEASURES BASED ON DISTANCES BETWEEN EXPERTS

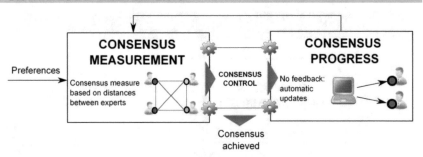

Fig. 2.10 Consensus models with automatic adjustments and different types of consensus measures

an expert's opinion and the collective opinion, i.e. the similarity between an individual assessment and the associated group assessment previously obtained by aggregating all the individuals' assessments. Those proximity values lying below a minimum proximity threshold are deemed as an indicator of assessments requiring to be brought closer to the rest of the group, i.e. they contribute to the overall consensus level being insufficient.

2. Producing and providing experts with feedback, indicating the previously identified experts how they should modify some of their assessments in order to bring them closer to the rest of the group. Most approaches in the literature undertake this task via direction rules, that specify (1) which expert should modify some of their opinions (2) which particular assessment(s) should (s)he revise, and (3) how to adjust such assessment(s). Examples of direction rule formats are:

- *Increase-decrease rules* [105]: These rules advise on the direction that the expert should follow in adjusting her/his assessments, without specifying the magnitude of the adjustment to be made. For instance,

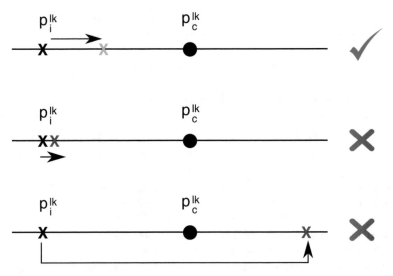

Fig. 2.11 Graphical example of appropriate and inappropriate adjustment of assessments based on a direction rule of the form: *Increase* p_i^{lk}

$$IF\ p_i^{lk} < p_c^{lk}\ THEN\ Increase\ p_i^{lk}$$

whilst an increase-decrease rule allows the expert for deciding the extent to which an assessment will be modified, it might foster situations of manipulative behaviors [38] if an expert decides to modify the opinion to a negligible degree (applying a very small, almost null increase/decrease on the assessment value) so as to preserve his individual interests rather than self-adapting to the collective concerns. This may particularly occur when a continuous numerical format is used for assessments, rather than e.g. discrete scales. Conversely, this direction rule format may also enable "random" patterns of strategic manipulative behavior [173] if an expert decides to apply a very drastic modification in an assessment, moving it "beyond the collective assessment" in the opposite direction. For the above reasons (graphically illustrated in Fig. 2.11), although usually effective in practice, increase-decrease rules shall be handled with care in practice.

- *Explicit adjustment rules* [96, 111]: This direction rule format provides more clear and specific guidelines not only on the direction to be followed (increase or decrease) but also on the magnitude of changes advised, e.g. "increase p_i^{lk} by 0.1" (for numerical assessments in the unit interval), or "decrease $p_i^{lk} = s_h$ into s_{h-1}" (for assessments expressed linguistically). Therefore, the responsibility of the expert would boil down to either accepting or declining the adjustment being proposed.
- *Adjustment rules within an allowed interval* [38]: Similar to increase-decrease rules, but in this case the allowed range of change is limited to an interval (e.g.

the semi-open interval between p_i^{lk} and p_c^{lk}), thereby preventing manipulation situations such as the one shown at the bottom example in Fig. 2.11. Example:

$$IF \; p_i^{lk} < p_c^{lk} \; THEN \; update \; \overline{p}_i^{lk} \in]p_i^{lk}, p_c^{lk}]$$

with $\overline{p}_i^{lk}$ used to denote the adjusted assessment for the next CRP round and $]a, b]$ a semi-open interval.

As opposed to feedback mechanisms involving an active involvement by experts throughout the CRP, automatic consensus building approaches [35, 160] rely on techniques that find an optimal adjustment of initial preferences to increase consensus. Optimization techniques such as constraint satisfaction, linear or non-linear programming models are typically employed to this end. Whilst the main goal in most cases is to minimize the sum of all distances between individual opinions and the collective opinion, further constraints are normally added e.g. finding an optimal consensus solution subject to minimum adjustments to preserve the initial views of participants as much as possible [10], preserving the consistency of opinions [150], or even finding an optimal setting of importance weights for experts such that the consensus level measured upon individual distances to the aggregated collective opinion is maximum [36]. A comprehensive compilation of consensus approaches based on automatic adjustments is provided by Dong and Xu in [35].

2.4.3 A Step-by-Step Example of Consensus Model

Once the main pillars inherent to any CRP have been presented, we now introduce an example of consensus model describing the necessary steps to find consensual solutions for a GDM. In particular, the example considers a GDM framework with the following characteristics:

1. The existing alternatives are evaluated "as a whole", without considering multiple decision criteria to assess them.
2. Experts feel comfortable to provide numerical pairwise comparisons among alternatives, thereby using additive preference relations.
3. A group leader or moderator assigns a priori different importance degrees λ_i, $i = 1 \ldots m$ to experts' preferences P_i, e.g. based on their level of expertise in the problem addressed.
4. The consensus level is measured based on similarities between pairs of experts, and a feedback mechanism is adopted for them to manually adjust their preferences in accordance with the advice received.

A consensus model incorporating the above characteristics is now described in detail.

(Step 1)   Gathering preferences

Each expert e_i in the group provides his/her opinions on alternative pairs (x_l, x_k) in X to the moderator, by means of an additive preference relation $P_i = (p_i^{lk})_{n \times n}$. It is assumed without losing generality that these experts provide reciprocally consistent assessments, i.e. expressing $p_i^{lk} \in [0, 1], l \neq k$, implies $p_i^{kl} = 1 - p_i^{lk}$.

(Step 2)   Consensus measurement

The agreement level in the group is calculated by undertaking the following sub-steps:

1. For each pair of experts $e_i, e_j, (i \neq j)$ a similarity matrix $SM_{ij} = (sm_{ij}^{lk})_{n \times n}$ is determined such that:

$$sm_{ij}^{lk} = 1 - |(p_i^{lk} - p_j^{lk})| \tag{2.12}$$

where $sm_{ij}^{lk} \in [0, 1]$ is the similarity degree between experts e_i and e_j in their assessments p_i^{lk}, p_j^{lk}.

2. A consensus matrix $CM = (cm^{lk})_{n \times n}$ is then yielded, taking into account the group's attitude, by aggregation of similarity matrices. Each element $cm^{lk}, l \neq q$, is computed as:

$$cm^{lk} = Attitude - OWA_w(SIM^{lk}, \vartheta, \varphi) \tag{2.13}$$

where $SIM^{lk} = \{sm_{12}^{lk}, \ldots, sm_{1m}^{lk}, sm_{23}^{lk}, \ldots, sm_{2m}^{lk}, \ldots, sm_{(m-1)m}^{lk}\}$ is the set of all pairs of experts' similarities in their opinion on (x_l, x_k). The attitude-OWA operator [101] is an extension of Yager's Ordered Weighted Averaging (OWA) operator, that integrates the group's attitude towards consensus predicated on two parameters ϑ and φ established a priori by the decision group. It extends the quantifier-guided OWA operators [170], such that $\vartheta, \varphi \in [0, 1]$ specify, respectively, the attitudinal character (degree of optimism) and amount of information integrated in the aggregation of pairwise similarities among experts. These two parameters, fixed by the decision group, provide enough information to define an associated instance of Attitude-OWA operator, as shown below. Attitude-OWA weights w_z for every element (pair of experts) are calculated as follows [170]:

$$w_z = Q\left(\frac{z}{n}\right) - Q\left(\frac{z-1}{n}\right) \tag{2.14}$$

with Q a regular increasing monotone quantifier defined by the following fuzzy membership function, $\alpha, \beta \in [0, 1]$:

$$Q(r) = \begin{cases} 0 & \text{if } r \leq \alpha, \\ \dfrac{r - \alpha}{\beta - \alpha} & \text{if } \alpha < r \leq \beta, \\ 1 & \text{if } r > \beta \end{cases} \qquad (2.15)$$

where $\alpha = 1 - \vartheta - \varphi/2$ and $\beta = \alpha + \varphi$. The interested reader is referred to [101] for further detail, mathematical and behavioral properties of the Attitude-OWA operator.

3. The consensus degrees are computed at three different levels:

   (a) Level of pairs of alternatives (cp^{lk}): Obtained from CM as $cp^{lk} = cm^{lk}$, $l, k \in \{1, \ldots, n\}, l \neq k$.
   (b) Level of alternatives (ca^{l}): The level of agreement on each alternative $x_l \in X$ is computed as:

$$ca^{l} = \frac{\sum_{k=1, k \neq l}^{n} cp^{lk}}{n - 1} \qquad (2.16)$$

   (c) Level of preference relation (overall consensus degree in the group). It is computed as:

$$CD = \frac{\sum_{l=1}^{n} ca^{l}}{n} \qquad (2.17)$$

(Step 3)   Consensus control

The consensus degree CD is compared with a consensus threshold $\mu \in [0, 1]$ established a priori. If $CD \geq \mu$, the group moves on to the alternatives selection phase; otherwise, they proceed to the feedback mechanism phase.

(Step 4)   Feedback Mechanism

If $CD < \mu$, then the group members need to modify their preferences to increase the level of mutual agreement in the following rounds. To do this, the feedback mechanism incorporated in this consensus model considers a number of actions:

1. Compute the collective preference P_c and proximity matrices for experts: the collective assessments p_c^{lk} are computed for each pair of alternatives by aggregating experts' preference relations, taking their importance degrees w_i into account:

$$p_c^{lk} = \phi_W(p_1^{lk}, p_2^{lk}, \ldots, p_m^{lk}) \qquad (2.18)$$

with ϕ_W the weighted averaging operator and $W = [\lambda_1, \ldots, \lambda_m]^T$. A proximity matrix $PP_i = (pp_i^{lk})_{n \times n}$ between each expert opinions and P_c can now be constructed for each expert: the proximity degrees pp_i^{lk} are obtained for each pair (x_l, x_k) as follows:

$$pp_i^{lk} = 1 - |(p_i^{lk} - p_c^{lk})| \qquad (2.19)$$

Proximity values between individual assessments and the collective assessment, are used to identify the furthest preferences from the collective opinion, which should be modified by some members.

2. Identify preferences to be changed (CC): Pairs of alternatives (x_l, x_k) whose consensus degrees ca^l and cp^{lk} remain insufficient, are identified:

$$CC = \{(x_l, x_k)|ca^l < CD \wedge cp^{lk} < CD\} \qquad (2.20)$$

Based on CC, experts who should change their opinion on each of these pairs, i.e. those individuals e_i whose assessment p_i^{lk} on the pair $(x_l, x_k) \in CC$ is furthest to p_c^{lk}, are identified. An average proximity $\overline{pp}^{lk}$ is calculated to do this:

$$\overline{pp}^{lk} = \phi(pp_1^{lk}, \ldots, pp_m^{lk}) \qquad (2.21)$$

As a result, individuals e_i whose $pp_i^{lk} < \overline{pp}^{lk}$ are advised to modify their assessment on (x_l, x_k).

3. Establish change directions: Several direction rules are applied to suggest the direction of changes proposed to experts, in order to increase the level of agreement in upcoming rounds of the CRP.

   - DIR.1: If $(p_i^{lk} - p_c^{lk}) < 0$, then e_i should *increase* his/her assessment on the pair of alternatives (x_l, x_k).
   - DIR.2: If $(p_i^{lk} - p_c^{lk}) > 0$, then e_i should *decrease* his/her assessment on the pair of alternatives (x_l, x_k).
   - DIR.3: If $(p_i^{lk} - p_c^{lk}) = 0$, then e_i should not modify his/her assessment on the pair of alternatives (x_l, x_k).

2.5   A Quick Overview of Multi-Criteria Decision Making Methods

Many of the existing GDM and consensus-driven methods involve the use of two or more evaluation attributes or criteria for assessing the available alternatives. For this reason, MCDM approaches are frequently integrated with the aforesaid methods, resulting into comprehensive MCGDM models involving both (1) individually supplied evaluations of alternatives under several criteria; (2) the necessity of fusing both individual assessments into a collective assessment, and aggregating collective evaluations across criteria into an overall assessment value for each alternative.

This section describes two popular MCDM methods with proven usefulness in systematically assessing alternatives in both individual and group decision scenarios [47]: AHP and TOPSIS. Numerous extensions and hybrid AHP-TOPSIS methods have been developed since the original two techniques were proposed.

Notwithstanding, for the sake of simplicity in discussion, here the two base versions of both techniques are described.

2.5.1 Analytic Hierarchy Process (AHP)

The AHP method was developed by Saaty [121–124] primarily inspired on the natural way humans act or respond when resolving a decision-making problem by decomposing it into sub-problems that are later aggregated to obtain a final recommendation. This method allows the expert to organize and express their judgments or feelings over the elements that form a decision-making problem (i.e. goal, evaluation criteria and alternatives) by a pairwise comparison to have an easy and controllable way to understand and solve the problem.

Basically, the AHP method guides the expert to decompose the decision problem through three main phases to generate the priorities over the set of alternatives:

1. Structure the decision problem as a hierarchical structure.
2. Construct a set of pairwise comparison matrices.
3. Calculate the priorities of the elements in the hierarchical structure.

In AHP all the considered set of alternatives and set of criteria of the decision problem must be arranged into a hierarchical structure which has a minimum of three levels. In the first level that is the top of the hierarchy, is placed the goal of the decision problem; in the second level are placed the set of criteria; and in the third level are placed the set of alternatives. Furthermore, this three-level structure can be extended with more levels to represent sub-criteria that break down into a complex decision problem. In Fig. 2.12 we show a simplified version of a hierarchical structure associated to a job selection decision-making problem presented in [125]. This is a decision making problem faced by a PhD student where he/she wants to know what would be the best job option for himself/herself when finishing his/her graduate studies. The first level in the structure is the desired goal of the PhD student for the decision problem, in this case to know which is the more suitable job. In the second level is the set of criteria $C = \{c_1, \ldots, c_z\}, z \geq 2$ defined to evaluate the alternatives in this problem and is composed by the reputation of the job, flexibility of the job, opportunities for professional grow, job security, and salary; in the last level is placed the set of alternatives $X = \{x_1, \ldots, x_n\}, n \geq 2$, encompassing two companies from the industry (a domestic and an international company), and two educational organizations (local college and a state university). We refer the reader to [125] for a more complex hierarchical structure of this problem.

This hierarchical structure lets us continue with the second phase which aims to make a pairwise comparison of the structured elements that are in a lower level with respect to the element they are connected to at an higher level. For example, in the hierarchical structure show in Fig. 2.12, for the elements in the second level (e.g. the set of criteria) the expert must give the importance or level of dominance, of

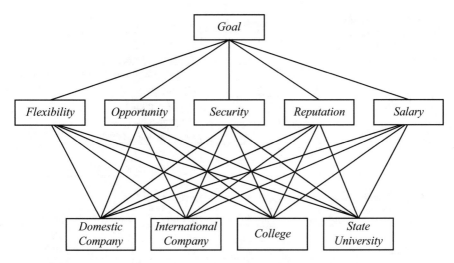

Fig. 2.12 Job selection problem hierarchical structure based on AHP

one element (criterion) over another element (criterion) regarding to the established goal. This could be seen as the experts answer to the question: how more important is the flexibility criterion over the opportunity criterion? The relative importance of one element over another is expressed on a scale proposed by Saaty of absolute numbers that are shown in Table 2.1. In this phase, the expert usually creates several comparison matrices depending on the number of elements in the hierarchical structure. For the example illustrated in Fig. 2.12, a total of 6 pairwise comparison matrices must be created: one matrix of the criteria over the goal and five matrices reflecting the comparisons of the set of alternatives over each criterion. In Table 2.2 we show the pairwise comparison matrix for the set of criteria over the established goal.

The third phase of AHP aims to calculate the priorities or scores of the elements in each comparison matrix to give meaning to the importance of each element in the decision-making problem. To calculate the priorities for the elements in a pairwise comparison matrix we can use the eigenvalue that has been widely used since the first AHP works or a variety of mathematical models as described in the literature [30, 66, 79]. These priorities are then used as weights to get an overall priority of the elements that are in the below levels until the global priorities of the alternatives, which are at the bottom level in the hierarchical, are calculated.

We refer the interested reader to [76, 120] for further discussions on the AHP technique, its various interpretations and variations.

Table 2.1 Importance
assessment scale (based on
Saaty's multiplicative scale)
and semantics of each value

Importance level	Verbal assessment
9	Extreme importance
8	Very, very strong
7	Very strong or demonstrated importance
6	Strong plus
5	Strong importance
4	Moderate plus
3	Moderate importance
2	Weak or slight
1	Equal importance (indifference)

Table 2.2 Pairwise comparison matrix for the six defined criteria

	Flexibility	Opportunities	Security	Reputation	Salary
Flexibility	1	1/4	1/6	1/4	1/8
Opportunities	4	1	1/3	3	1/7
Security	6	3	1	4	1/2
Reputation	4	1/3	1/4	1	1/7
Salary	8	7	2	7	1

2.5.2   Technique for Order of Preference by Similarity to Ideal Solution (TOPSIS)

TOPSIS is a MCDM method that has been widely accepted and has received much interest from researchers and practitioners to solve real decision-making problems from different areas such as: supply chain management and logistics, design, engineering and manufacturing systems, business and marketing management, health, safety and environment management, human resources management, energy management, chemical engineering, water resources management, among others [7]. This method was first presented in [65] as a simple model to solve the multicriteria ranking problematic and have had several contributions and improvements since it conception. We refer the interested reader to [42, 133] for relevant recent developments on TOPSIS-based techniques.

The core of TOPSIS is based on the concept of compromise programming which aims to set an ideal solution as a reference point according to the experts' preferences and then seek for those solutions whose attributes are closer to an ideal solutions attributes [183]. Basically, TOPSIS defines two fictitious alternatives known as positive and negative ideal solution. The positive solution reflects the best alternatives assessment on each criterion while the negative ideal solution reflects the worst ones. Then, a geometric distance (usually the Euclidean distance) of each decision alternative is calculated to both, positive and negative ideal solutions. The alternative that is closer to the positive ideal solution and farthest from the negative ideal point is considered the best alternative among the set X of alternatives, therefore, this distance is used as a score to rank the set of alternatives.

The original TOPSIS method is usually given by five main steps and can be described as follows:

Assume the existence of a set X of $n \geq 2$ *alternatives* or possible solutions, a set C of $z \geq 2$ evaluation criteria, and a decision matrix M that reflects the assessment of each alternative on each criterion. Also assume that the decision maker or expert has given his preferences by defining a set of weights $W = \{w_1, \ldots, w_z\}$ reflecting the relative importance of each criterion. We suppose without loss of generality that $(\sum_{q=1}^{z} w_q = 1)$

1. The first step is to normalize the values or assessments on alternatives under each criterion from the different scales. For this, it is created and calculated a normalized decision matrix $R = (r^{lq})_{n \times z}$ from the original decision matrix M by:

$$r^{lq} = \frac{x_{lq}}{\sqrt{\sum_{l=1}^{n} x_{lq}^2}}, l = 1, \ldots, n, q = 1, \ldots, z \qquad (2.22)$$

2. The second step is to aggregate the total importance of each alternatives on each criterion by calculating a weighted normalized decision matrix $V = (v^{lq})_{n \times z}$ from the normalized decision matrix R as:

$$v^{lq} = w_q v^{lq}, l = 1, \ldots, n, q = 1, \ldots, z \qquad (2.23)$$

3. The third step is calculate, from the weighted values, the fictitious positive ideal solution A^+ and negative ideal solution A^- alternatives as:

$$A^+ = \{v_1^+, \ldots, v_z^+\}, where \quad v_q^+ = \max_l(v_{lq}), l = 1, \ldots, n, q = 1, \ldots, z \qquad (2.24)$$

$$A^- = \{v_1^-, \ldots, v_z^-\}, where \quad v_q^- = \min_l(v_{lq}), l = 1, \ldots, n, q = 1, \ldots, z \qquad (2.25)$$

Please note, that here we are assuming that all the criteria are to be maximized, thus the higher the value on each criterion the better it is deemed.

4. The fourth step is to calculate, for each alternative, their distance d_l^+ to the positive ideal solution A^+ and the distance d_l^+ to the negative ideal solution A^- by:

$$d_l^+ = \sqrt{\sum_{q=1}^{z} (v_q^+ - v_{lq})^2}, l = 1, \ldots, n \qquad (2.26)$$

$$d_l^- = \sqrt{\sum_{q=1}^{z}(v_q^- - v_{lq})^2}, l = 1, \ldots, n \qquad (2.27)$$

5. Finally, the relative closeness coefficient C_l of each action to the ideal solution is calculated by:

$$C_l = \frac{d_l^-}{d_l^+ + d_l^-}, l = 1, \ldots, n \qquad (2.28)$$

The coefficient is used to rank the alternatives according to the positive ideal solution in a descending order of the distance value. It should be noted that if an alternative's coefficient is near one then it means that the alternative is closer to the positive ideal solution, whereas is the alternative's coefficient is near zero then it means that the alternative is closer to the negative ideal solution.

Chapter 3
Scaling Things Up: Large Group Decision Making (LGDM)

Abstract What is a Large Group Decision Making problem? What differentiates them from the conventional Group Decision Making problems and approaches introduced in the previous chapter, and what are the added complexities of supporting high-quality decisions to be made by large groups? The present chapter aims at introducing and contextualizing this relatively new area of research, highlighting its main limitations of challenges and discussing some of its newly related disciplines, as witnessed in recent research.

3.1 From Small to Large Decision Groups

Classically, GDM problems taking place in most real-world settings have been implemented at a strategic level, where decisions are often made by a small number of people, e.g. the executive team or advisory board in an organization. However, the recent proliferation of new technological and societal paradigms are making possible—and increasingly advocated—the participation of a large body of individuals in the decision making processes. In large organizations, for instance, the last two decades have witnessed remarkable changes in the context where collective decision-making processes occur [22]. The emergence of global markets, the proliferation of multinational firms and the increasingly globalist view of society [131] make it necessary for directive team members spread around the world (often under very different time zones) to make strategic decisions jointly. Other obvious examples of novel paradigms demanding collective decisions at scale include social media platforms [128], e-democracy systems [25], e-marketplaces for group and corporate shopping, group recommender systems and crowd-funding platforms [43, 63], amongst others.

As a result, Large Group Decision Making (LGDM) problems, in which a large group of experts take part in a decision problem jointly, has become an important sub-area of study within the research community devoted to GDM and consensus building. LGDM scenarios are actively investigated nowadays in both recent and ongoing works in the area, as well as reaching out to other disciplines. But, what differentiates a classical (small group) GDM problem from a LGDM problem?

© The Author(s), under exclusive licence to Springer Nature Switzerland AG 2018
I. Palomares Carrascosa, *Large Group Decision Making*, SpringerBriefs
in Computer Science, https://doi.org/10.1007/978-3-030-01027-0_3

Whilst there is no clear answer to this question, some authors argue that groups formed by e.g. 11 or more members or at least 20 members shall be considered as large groups. By looking at the extant literature, we postulate however that a LGDM problem differs from a conventional GDM problem not only in group size, but rather in its inherent *higher complexity* to make an accepted collective decision. Such complexity can be influenced by both group size and other factors, e.g. group diversity. Accordingly, a possible approach for defining a LGDM problem is now introduced from the author's perspective.

Definition 3.1 A Large-Group Decision Making problem (LGDM) is a situation involving between several tens and thousands of participants with *diversity* in background, expertise level, behavior, attitudes and possibly conflicting interests/viewpoints, who must make a collective and acceptable decision pertaining a relevant problem to all of them.

3.2   Limitations and Challenges

Despite the large amount of models and approaches that have been proposed by a variety of authors to support CRPs in GDM problems, they have normally focused on problems where a small number of experts take part. The research results obtained in this field of study until a few years ago, were not sufficient when dealing with large-scale GDM problems: new difficulties and challenges have been reported to arise [108], which require further study for the improvement of CRPs in which a large number of experts are involved. Some of these limitations and challenges are described below:

- **Developing scalable and distributed decision support architectures**: Reliable, fault-tolerant and highly scalable group decision processes [64] become a paramount aspect to consider when the size of the group increases to tens, hundreds or even thousands of participants. Clearly, without appropriate GDSS architectures, undertaking such processes at a large scale may become a daunting or sometimes impossible task. Efficient and effective decision support architectures must therefore be implemented to (1) accommodate and deal with large amounts of decision information in a time-sensitive manner (sometimes this information is not only limited to experts' preferences but it may also incorporate other dimensions of relevant data) [134, 153, 163], (2) enable ubiquitous LGDM processes in which the participants may be geographically distributed [24, 100], or (3) facilitate coordination and innumerable flows of communication and exchanges of ideas between (lots of) participants, in scenarios where the decisions are made through pairwise or small-group interactions [48, 64].
- **Identifying and managing participants' behaviors**: Making consensual group decisions implies that experts must discuss and modify their initial preferences, moving their opinions closer to each other towards a collective solution which

satisfies the whole group to a reasonable extent, even though this sometimes implies accommodating changes in their initial individual interests. Naturally, the assumption that *all* experts may always accommodate the group interests in detriment of their individual interests is neither realistic nor common in practical situations. Instead, some participants are more prone than others to bringing their own opinions closer to the other members in the group, in other words, some experts are more cooperative than others in GDM problems involving CRPs. The problem of dealing experts who present a non-cooperative behavior in CRPs, because they are reluctant to modifying their initial positions and making them closer to the rest of experts in the group, has quickly become an important aspect of research in the LGDM community [22, 36]. The presence of individuals—or subgroups of them—whose behavior does not contribute to the achievement of consensus, is particularly frequent (and thus more difficult to manage) in LGDM problems [108]: in large groups, it is more common to encounter not only uncooperative individuals, but also subgroups or coalitions of experts with similar interests who jointly attempt to manipulate the CRP and deviate the overall outcomes of the LGDM process in their favor. Put another way, some of these coalitions may decide not to modify their preferences, or even they might modify them against the rest of experts' positions coordinately, with the aim of introducing a bias in the collective opinion that benefits their own interests. These behaviors would affect the CRP performance negatively, since they might significantly hinder the process of finding a collective agreement.

- **Defining cost-effective approaches for LGDM and building consensus**: By analyzing existing proposals of consensus models and CSS in the literature, a notable shortcoming found in feedback-based approaches requiring human intervention, is the necessity of constant supervision of preferences by human experts throughout several CRP rounds. This human supervision requires investing a substantial amount of time in revising and modifying preferences, therefore it may often cause an excessive time cost and, in some cases, the eventual loss of motivation and interest of experts in the problem being addressed [104]. In order to tackle this added difficulty in large-group CRPs in situations where individuals should still actively participate to some extent, recent studies [104, 111] have suggested the definition of semi-supervised consensus models, whereby a compromise solution is implemented lying between feedback-based and automatic approaches for consensus building. Depending on the characteristics of each generated advice, the expert's supervision and acceptance of such feedback is requested or not: where the proposed change on her/his opinions is not significant, the opinions are adjusted automatically instead. Because human supervision is not eliminated completely, the sovereignty of experts' is preserved to some degree, unlike it occurs in some proposals of automatic consensus models in the literature, in which this sovereignty is "eliminated" and once experts provide their initial opinions they are no longer in control of any changes occurring on such opinions [160].

Another notable approach to ensure cost-effective LGDM processes, which has arguably given rise to the most extended trend of research within LGDM, is that of subdividing the large group into multiple subgroups, each of which presents a smaller and more manageable size. This way, the LGDM processes can be separated into multiple smaller GDM processes taking place in parallel, after which decision results produced by each subgroup are aggregated into a final result. Most subgroup-driven approaches [105, 152, 194] employ clustering techniques based on the similarity among experts' preferences. Furthermore, adequate strategies for aggregating and weighting information at individual level, subgroup level and large-group level must be likewise considered.

- **Monitoring the progress of the LGDM process**: A visual analysis of the LGDM processes, particularly those involving a CRP, would be desirable in many scenarios to have an insight of the positions of participants with respect to each other, as well as to facilitate the rapid detection of majority and minority groups, non-cooperating experts and the overall evolution of experts' preferences across a CRP [107, 130]. For CRPs in which few experts take part, these monitoring tasks can be easily undertaken by means of supporting tools based on textual or numerical information [15, 71]. However, the large amount of information utilized in large-scale GDM problems necessitates the use of new tools capable of providing more easily interpretable information about the state of the LGDM process from beginning to end. To overcome this challenge, recent research interest has shifted towards visual monitoring tools to be used along with GDSS to further support large groups or analysts, for example by visualizing the preferences of participants in a low-dimensional space [103, 107].

- **Implementing adequate strategies for weighting and aggregating decision information**: The process of aggregating decision information or assigning reliable weights to participants and/or evaluation criteria, also deserves further attention under the presence of large groups. For example, novel and effective weighting and aggregation strategies should be studied in situations where both majority subgroups and minority subgroups coexist, namely to ensure that the results of weighting/fusion process do not end up "accidentally" eliminating any valuable information from underrepresented minority opinions [165]. Large groups divided into clusters [152], the presence of different behaviors [36], and the existence of social relationships information between participants [192], constitute other examples of relevant factors that must be carefully considered when defining the most suitable aggregation and weighting strategies in LGDM methodologies. In essence, simple strategies that provide desirable results in small-group problems, e.g. a weighted arithmetic mean-based aggregation of individual preferences, or the subjective determination of experts weights, are in many cases no longer suitable to find the best solution in a LGDM problem where participants are *highly diverse* and additional meaningful information describing such diversity is available. Figure 3.1 depicts an example of MCGDM scenario where—by considering exclusively the preferences provided by an

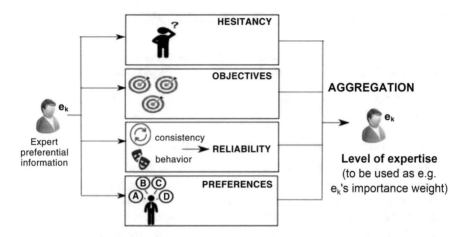

Fig. 3.1 Example scenario proposing the analysis of multiple expertise indicators upon an expert's preferences

expert as decision matrices with uncertain assessments—different attributes can be considered as relevant indicators of her/his expertise level in a LGDM problem.

- **Incorporating additional knowledge and data sources**: Large decision groups are rarely formed by experts who are socially isolated from each other. Instead, it is usual that different social relationships with varying strength (trust, distrust, reputation acquired, etc.) exist between them. For this reason, recent LGDM and consensus models are incorporating social information to have a better understanding of the participants' background, their motivations and attitude in the decision making process [130, 192]. Other scenarios may involve the existence of Internet user profiles associated to the LGDM participants, or data describing their participation history in previous problems, their behavior and satisfaction with previous decision outcomes, etc. There is an still somewhat untapped potential in exploiting and acquiring insight from such data sources, which to date have been scarcely explored in the GDM and LGDM literature, as it may help defining improved and highly-tailored LGDM and CRP models that take as much relevant information as possible into account. The marriage between this area of study and other related ones, such as group recommender systems for large collectives [43] and opinion dynamics/formation [40], is another challenge being currently tackled.
- **Deploying and validating new models in real problems**: Despite the growing interest in investigating LGDM models and methods, by exhaustively analyzing the existing literature it is clear that very few works to date have attempted to deploy, test and apply such models in real situations. Aspects requiring further study include, for instance, the deployment of a larger number of methods into real Web or mobile GDSS, or defining and implementing robust validation approaches e.g. to measure user's satisfaction with the decisions made.

3.3   Summary of Research Trends on LGDM

As pointed out in various contributions to the LGDM literature [189, 193], related works in the field are usually classified into various categories or *trends*:

1. Subgroup clustering approaches.
2. LGDM methods.
3. Consensus approaches for LGDM.
4. Large Group Decision Support Systems (LGDSS).

However, based on the challenges and limitations described in the previous section, and in order to better represent the extensive recent literature on LGDM (which is reviewed Chap. 4), this book provides an extended categorization for existing LGDM research. The motivation for an extended classification of LGDM studies is motivated by the following difficulties, which have been encountered in the above fourfold classification in our attempts to exhaustively categorize all the existing approaches to date:

- *Difficulty to categorize some existing studies*: There exist some contributions to LGDM that do not significantly fit in neither of the above four categories, e.g. because they focus on theoretical and/or interdisciplinary studies.
- *Categorization interpretation*: It should not be considered that the existing studies focus their contribution on *one and only one* of the above categories, largely because they may present notable contributions related to several of them (e.g. an innovative consensus model for LGDM that also introduces novel subgroup clustering techniques).
- *Behavior management as a standalone trend*: Although all LGDM methodologies for handling experts' behavior are classified as consensus approaches for LGDM, the contribution in several of them concentrates almost exclusively on the modeling of management of these behaviors during the CRP, rather than defining novel aspects in the CRP itself (e.g. defining a new consensus measure or an innovative feedback mechanism). Therefore, it would be reasonable to consider behavior management as a sub-trend of consensus approaches within the area of LGDM.
- *Unclear characterization of 'LGDM methods' category*: Despite clearly being one of the four core categories for classifying existing LGDM works, there is no clear explanation in the literature on what specific aspects do LGDM methods investigate.

In order to overcome these limitations, an extended categorization is proposed and presented in this book (next chapter), under the following considerations:

- A new category entitled *Theoretical and Interdisciplinary Studies*, is introduced to cover existing contributions that otherwise are difficult to allocate into one of the four frequently considered trends.
- We herein propose a single categorization, pointing out that some of the reviewed literature works consider contributions under two or more trends: the "primary trend" is thus assigned based on the primary contribution made by the author(s).

Each work is reviewed in the section that corresponds to its primary trend, but where a secondary trend(s) may exist for some works, these are likewise mentioned in the text.

- *Behavior modeling and management* is introduced as an additional trend, conceptually deemed as a subcategory within *consensus approaches for LGDM*.
- A clear and succinct description is provided for each of the newly considered LGDM trends in the literature review. This particularly aids in better understanding the scope and aspects studied under the *LGDM methods* trend.

In short, Chap. 4 introduces a novel extended classification of extant LGDM literature, and provides a detailed literature review of existing LGDM studies divided into several categories. Chapter 5 focuses on works describing GDSS implementations for large groups, and briefly discusses the most popular application domains for putting LGDM approaches into practice.

3.4   Related Disciplines to LGDM

The proliferation of LGDM approaches in recent years has exposed the potential relationship between this area of research and other disciplines (Fig. 3.2). Most conventional GDM models have exclusively relied on principles from decision making under uncertainty and consensus support, with proven effectiveness without

Large-Group Decision Making (LGDM): A large number of experts with diverse *backgrounds, expertise level, behavior* and possibly conflicting interests/viewpoints.

   I.   COMPLEXITY: Necessity of <u>scalable</u> decision support tools and approaches
   II. FAIRNESS: Managing non-cooperative <u>behaviors</u> of individuals or subgroups
   III. INSIGHT: Monitoring the LGDM process
   IV. PROCESS: Optimally <u>aggregating and weighting</u> lots of (uncertain) decision information
   V.   DATA: Considering new sources of relevant decision Information (e.g. social)
   VI. CONSENSUS: Temporal <u>cost</u> of making consensual decisions

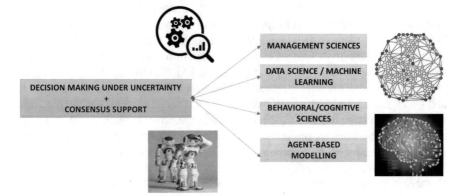

Fig. 3.2 Summary of LGDM challenges and some of its main related disciplines

the need for considering other areas of knowledge. However, these principles are no longer sufficient to adequately handle large and highly diverse groups, as well as new sources of relevant decision information, thus it often becomes necessary to adopt principles from other—previously not explored—disciplines. Below we enumerate some of these disciplines, justifying their relationship with LGDM.

3.4.1 Cognitive and Behavioral Science (Psychology)

Various studies [5, 48] suggest an important relationship between group decisions and the psychological attitudes/behavior exhibited by their participating members. For example, Back et al. conducted in [5] a study on the post-decision opinion consolidation in large group members, from a neurosciences and cognitive sciences perspective. By investigating decision mechanisms predicated on in-group and out-group authority, the study concludes two important findings: (1) a decision-making perceived by participants as *fair* promotes a higher degree of acceptance of the decision made; and (2) a participant with individual opinions in discordance with the decision outcome, shall further increase her/his sense of attractiveness towards her/his most preferred alternatives in the future. More interdisciplinary approaches to LGDM adopting cognitive and behavior principles are reviewed in Chap. 4, Sect. 4.6.

3.4.2 Management and Social Sciences

Unsurprisingly, many popular principles adopted in MCDM and GDM (e.g. multi-attribute utility theory or the *Theory of Games and Economic Behavior* [148]) have their roots in decision analysis, which in turn is deemed as a subfield of management sciences [132]. With its origins traced back to operations research, management sciences are regarded as a broad interdisciplinary area of study of problem solving and decision making in human organizations of any nature [6]. It usually relies on methods from mathematics, statistics, economics and engineering. With the proliferation of LGDM models for supporting strategic decisions jointly made across departments or organizations [80, 84], it is sensible to start paying more attention towards models explaining how such decision processes take place, and which managerial factors are involved. From a social sciences perspective, it is noteworthy the large body of classic GDM and consensus research based on *social choice theory* methods for preference (vote) aggregation, which combine ideas from voting theory and welfare economics [1, 3, 46]. It subtly differs from GDM in that social choice methods do not necessarily assume that people in a group are priori committed to make a decision together. Social choice models and methods have also evolved towards large groups or societies, as witnessed in several LGDM works

[116, 117, 135]. On another note, management problems across myriad sub-domains (disaster management, administration, medical, policy-making, etc.) undeniably constitute the most frequent real-world application of LGDM research.

3.4.3   Data Science, Machine Learning and Artificial Intelligence

There is little doubt about the sheer expansion of these disciplines [44, 119] to almost every aspect of our lives over the last decade, but they also hold an enormous latent value to support—or be supported by—decision making process led by increasingly large groups [33, 64, 194]. Humongous decision groups, in essence, should be understood as humongous relevant data to be taken account in the decision making process itself. Even in strategic decisions to be ultimately made by small groups, opinion information from "the crowd" may still constitute a valuable input when it is considered. For example, by gathering and analyzing data from a large body of customers, its extracted knowledge can be effectively integrated into a strategic GDM problem for the development of a new product [29, 174]. On the other hand, well-known clustering techniques have already been applied to numerous LGDM models to analyze a large group and subdivide it into subgroups based on the similarity between their members [105, 194], and dimensionality reduction strategies such as Principal Component Analysis (PCA) have been also adopted to reduce the size-complexity of MCLGDM problems [85]. Therefore, it would be unsurprising to witness yet additional links to come between LGDM and other data science and machine learning techniques, such as classification, anomaly detection, pattern recognition, etc. From an Artificial Intelligence perspective, the rapid developments in agent-based modeling, autonomous intelligent systems and complex systems simulation across various domains [17, 23, 110] are turning the mutual collaboration and collective decision making involving both human and "artificial" participants into a reality. Autonomous vehicles and autonomous entities in smart cities or smart environments capable of reasoning and expressing their own preferential information, are examples of such new types of non-human participants in collective decisions. Thus, some efforts shall be required in developing new LGDM and consensus models for transparently dealing with large-scale scenarios involving (totally or partly) artificial intelligent participants in them [33, 77], as well as seamlessly integrating their distinct representations of the information and types of uncertainty exhibited (e.g. fuzzy and vague uncertainty shown by human participants *vs* probabilistic uncertainty exhibited by an autonomous device).

Chapter 4
LGDM Approaches and Models: A Literature Review

Abstract Once the foundations and main considerations for Large Group Decision Making have been set out, this chapter provides a comprehensive literature review of most of its related works in the scientific literature to date. The reviewed studies are categorized into subgroup clustering methods, Large Group Decision Making methods (for preference aggregation and weighting), consensus approaches, behavior management and modeling methodologies, and theory/interdisciplinary approaches.

4.1 Considerations and Organization of the Literature Review

Further to the extended categorization of LGDM research in close accordance with the limitations and considerations highlighted in the previous chapter, a total of six LGDM trends—subdivided into themes—are defined to underpin the literature review presented in this chapter:

- *Subgroup Clustering*: This trend relates to LGDM approaches characterized by dividing the large decision group into smaller—and usually more manageable—subgroups or clusters. A common aim in most subgroup clustering approaches is to reduce the cost and complexity of undertaking decision making (and consensus building) processes at large scale, such that these processes are partly or completely conducted at subgroup level in parallel. Another typical application of subgroup clustering is to find common opinion patterns across a large group such as subgroups with highly similar opinions to each other, in which one of its members is identified as a *spokesperson* who represents the whole subgroup. As the trend name suggests, some approaches based on subgroup clustering apply a clustering algorithm predicated for instance on preference similarity, for dividing

---

The original version of this chapter was revised. A correction to this chapter is available at https://doi.org/10.1007/978-3-030-01027-0_7

the large group into subgroups of "likeminded" participants with similar opinions or interests; whilst other approaches assume the existence a priori of multiple decision groups, e.g. different departments in an organization. Research related to this trend is surveyed in Sect. 4.2.

- *LGDM Methods*: Approaches under this trend focus on studying how the underlying computational processes in conventional selection-based methods for GDM and MCGDM methods can be translated into large group settings. Accordingly, aspects investigated by LGDM methods include: (1) the modeling and aggregation of large amounts of preferential information in an efficient and accurate manner; and (2) methodologies for weighting decision information such as assigning importance to each expert and evaluation criterion. Research related to this trend is surveyed in Sect. 4.3.

- *Consensus Approaches for LGDM*: This trend investigates the challenges and difficulties of building a high agreement level and making consensual decisions in large groups. Therefore, approaches under this category are assumed to incorporate a consensus phase prior to the aggregation-exploitation phase to select or rank alternatives. Examples of common aspects investigated under this trend are the study of the efficiency or convergence to achieve a target consensus level in a large group, and the definition of enhanced feedback mechanisms or preference adjustment methods to increase the agreement level. Research related to this trend is surveyed in Sect. 4.4.

- *Behavior Management and Modeling*: This trend considers LGDM situations where some participants adopt different behaviors towards the process of finding a highly accepted collective decision. Its related works investigate the process of describing, identifying and dealing with uncooperative behaviors so as to minimize the risk and negative effects of making a biased decision or a wrong decision as a result of a manipulated process by non cooperating individuals. Conceptually, this trend can be deemed as a subcategory of *consensus approaches for LGDM* since all behavior management and modeling approaches investigated in LGDM and its extensions assume that a CRP is conducted, whereby some participants may exhibit different patterns of cooperation during the iterative process of building consensus. Nevertheless, owing to the number of recent works and contributions specifically focused on managing experts' behaviors in CRPs rather than investigating the CRP itself, this trend is surveyed separately in Sect. 4.5.

- *Theory and Interdisciplinary Studies*: The reviewed studies under this trend are mainly focused on discussing theoretical aspects of decision making with large groups and/or putting a high emphasis on other relevant disciplines that have not been sufficiently investigated by most authors in the other LGDM trends. Research related to this trend is surveyed in Sect. 4.6.

- *Large Group Decision Support Systems (LGDSS)*: This trend focuses on real-world implementations of LGDM approaches and methodologies into computer-based, Web-based or mobile-based DSS. Most extant LGDSS are primarily aimed at enabling ubiquitous large-group decision processes in which participants are geographically separated from each other or they may even take part in the collective decision problem at distinct times. Due to their mostly practical

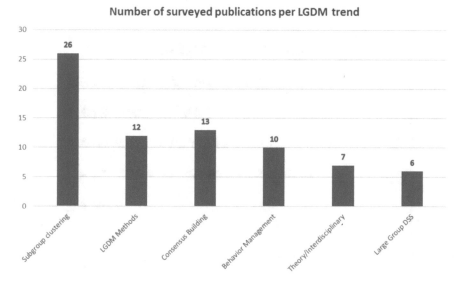

Fig. 4.1 Number of reviewed LGDM works per research trend

and application-oriented perspective, approaches under this trend are surveyed in a separate chapter. In particular, the architecture and main characteristics of some representative LGDSS implementations are presented in Chap. 5 of this book.

Before moving into the detailed survey, it is worth depicting some basic statistics about the total **74 LGDM publications reviewed** in the book. Figure 4.1 shows the number of LGDM publications surveyed for each of the above identified six trends. Clearly, subgroup clustering approaches represent the predominant research trend at present, whereas there is an apparent shortage of research focused on LGDSS implementations. Special interest deserves the plot shown in Fig. 4.2, which represents the number of LGDM works per year of publication (as of May 2018). There is an undeniably growing trend into LGDM research over the last few years, particularly from 2014 onwards, a year in which several pioneering LGDM solutions focused on supporting large-scale CRPs, were published [108]. This consolidates the state of LGDM as a rather young but rapidly expanding area of study. It can be also witnessed that some precursor studies for the state-of-the-art research date back to several decades ago.

4.2 Subgroup Clustering

Surveyed works on subgroup clustering are summarized in Table 4.1 and subdivided in the following five themes (Table 4.2):

- Early efforts on subgroup clustering in LGDM.
- Clustering methods for MCLGDM and complex MCLGDM.

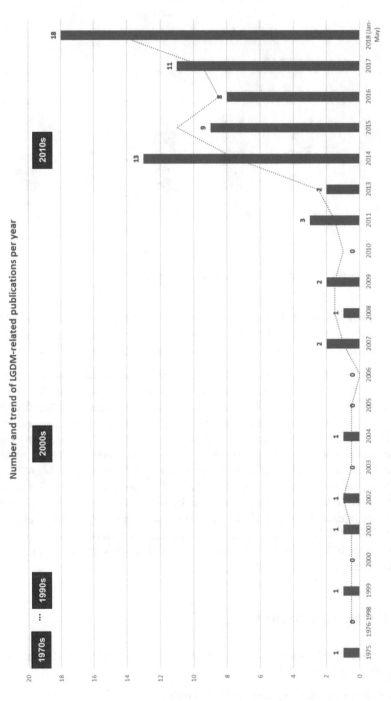

Fig. 4.2 Number and temporal trend followed by reviewed LGDM works per year of publication

Table 4.1 Summary of surveyed works on subgroup clustering

Theme	Representative publications (listed by theme alphabetically)
Early efforts on subgroup clustering in LGDM	Alonso et al. [2]; Bolloju [13]; Cole and Sage [32]; Zahir [182]
Clustering methods for MCLGDM and complex MCLGDM	Liu et al. [83, 86, 89]; Srdjevic [135]; Wang [149]
Clustering large groups in emergency and risk situations	Cai et al. [19]; Xu et al. [161–163]
Clustering methods under fuzziness	Tapia-Rosero et al. [138–141]; Wu and Liu [152]; Wu et al. [153, 154]
Other notable contributions to subgroup clustering in LGDM	Liu et al. [80, 87]; Soto et al. [134]; Xu et al. [159, 168]; Zhu et al. [194]

Table 4.2 Examples of clustering algorithms used in different LGDM and consensus models reviewed in this chapter

Clustering algorithm	Related LGDM works
Partitional divisive clustering	Cai et al. [19]; Liu et al. [80]; Xu et al. [159, 161, 162]; Zahir [182]
Partitional clustering (e.g. K-means)	Alonso et al. [2]; Bolloju [13]
Hierarchical clustering (e.g. HAC)	Tapia Rosero et al. [138–141]; Zhu et al. [194]
Fuzzy clustering (e.g. FCM, FEC)	Wang [149]; Wu et al. [152–154]; Xu et al. [163]
Discriminant analysis (e.g. DEA-DA)	Liu et al. [83, 86, 89]
Self-organizing maps (SOMs)	Xu et al. [168]
Not specified/cluster a priori	Cole and Sage [32]; Liu et al. [87]; Soto et al. [134]; Srdjevic [135]

- Clustering large groups in emergency and risk situations.
- Clustering methods under fuzziness.
- Other notable contributions to subgroup clustering in LGDM.

4.2.1   Early Efforts on Subgroup Clustering in LGDM

In the middle 1970s, Cole and Sage in [32] devised arguably the main precursor study for later subgroup clustering methods to support large group decisions. The authors targeted large-scale voting problems (e.g. choosing presidential nominees

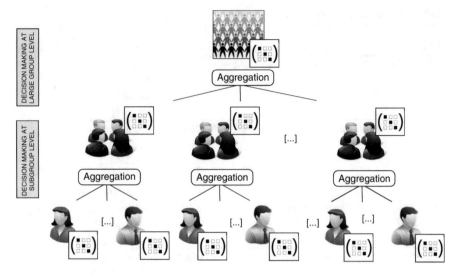

Fig. 4.3 General scheme of the two-stage preference aggregation and decision making process for large groups divided into multiple subgroups

for two major US parties) taking place in a decentralized manner, where the large group (society) is assumed to be subdivided into K subgroups with different number of individuals per subgroup. Assuming that the opinions are expressed as non-strict preference orderings, this study if pioneering in dividing a large-group decision making process into two broad stages: (1) decision making at subgroup level, focused on obtaining a subgroup preference ordering by aggregating individual preference orderings, and (2) decision making for the whole large group, aimed at yielding a "society preference ordering". The fusion processes at both levels are made by applying restricted social welfare functions. The study also defines an extension of Arrow's original work [3], addressing the research question on how to weigh preferences of unequal subgroups, i.e. groups with difference size and degree of importance from each other. They concluded that the majority rule in these situations can lead to intransitive results, in other words, it may cause the so-called "paradox of voting".

Remark 4.1 The two-stage aggregation and decision making process outlined above has been largely adopted in numerous subgroup clustering methods and LGDM approaches in general, for different types of preferences and decision frameworks. Its general scheme is depicted in Fig. 4.3.

In [182], Zahir presented a pioneering contribution to clustering into subgroups based on a similarity measure between preference vectors expressed numerically. Zahir's subgroup clustering approach is devised and applied within an AHP framework, and at the same time the author investigated the cohesiveness of the clusters created. The preference similarity measure consists in the scalar product between numerical preference vectors (given by the cosine of the angle between them). In addition, two measures for coherence are discussed for different patterns of information associated to individuals belonging to the same subgroup. The AHP framework with subgroup partitioning is showcased for evaluating the balance between economic and environmental impact of a plan involving the construction of a processing plant.

In [13], Bolloju established an AHP-based framework for modeling large group decentralized decision problems. In decentralized problems, traditional preference aggregation techniques are not feasible to fuse individual opinions, since the group homogeneity of aggregated preferences is more likely to be violated. Owing to the group homogeneity being deemed an indispensable condition for classical decision models such as AHP, Bolloju introduced a clustering approach to identify homogeneous subgroups of experts based on the similarity between their opinions, represented as utility functions previously unified based on varying subgroup factors. AHP aggregation, namely via the WGM and WAM operators, can then be applied at cluster level to aggregate individual preferences, and subsequently to aggregate subgroup preferences. The AHP-based clustering framework is applied to solve operational decision making problems involving 70 employees, on loan evaluation and risk analysis.

Alonso et al. introduced in [2] one of the earliest approaches for LGDM based on fuzzy preference modeling and aggregation. The authors argued in their work that decisions involving a large number of individuals are increasingly common e.g. in online communities about certain topics of interest (for instance, to collectively choose a venue and date for an upcoming large-group meeting). Before initiating a CRP, and based on a K-means clustering method, the large decision group E is "simplified" into a smaller group of selected experts or *spokespersons* $E_S \subset E$, in a manner that preserves the diversity among the original individual opinions as much as possible. The remaining experts in $E - E_S$ are allowed to express their degrees of trust in the selected spokespersons, thereby constituting a trust network. Figure 4.4 shows the process of identifying a small group of spokespersons and establishing a trust network upon an initial large group. Importantly, the model assumes that experts know each other and are able to provide comprehensive trust information about their peers accordingly. A CRP is then applied to seek a highly accepted group decision. Finally, in the selection process, an instance of the Induced OWA (IOWA) operator [172] is used to fuse individual preferences of spokespersons, predicated on the overall trust put on each one of them as the inducing variable of the group.

4.2.2 Clustering Methods for MCLGDM and Complex MCLGDM

In [135], Srdjevic studied MCLGDM problems with a priori constructed clusters. Similarly to Bolloju's study [13], the author hypothetized that most MCGDM

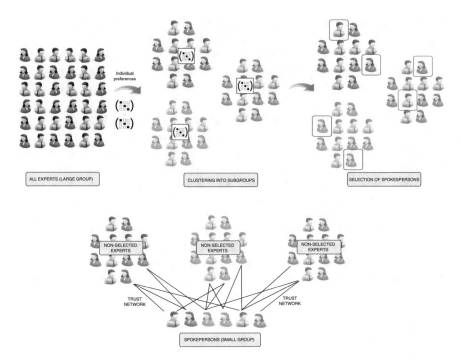

Fig. 4.4 Selection of spokespersons and creation of a trust network in the LGDM model by Alonso et al. [2]

methods rely on the homogeneity principle, which in large groups may no longer apply, and even within subgroups might sometimes pose a tedious preference elicitation process. Accordingly, similarity-based clustering is advocated to overcome this challenge. Assuming previously identified subgroups and their subgroup preferences derived a priori, Srdjevic focused on investigating the advantages and disadvantages of applying subgroup clustering methods in combination with (1) a MCDM framework based on AHP (with WAM and WGM as aggregation methods), (2) a social choice framework (with the Hare System as aggregation method), or (3) a mixture or both. Preferences are elicited as preference relations in Saaty multiplicative scale or Lootsma scale, i.e. ($\{-8, -6, \ldots, 0, \ldots, 6, 8\}$) [92].

The clustering of linguistic preferences and highly uncertain subgroup clustering problems have also been investigated over the last years. For instance, Wang developed in [149] a method for MCLGDM under both uncertain linguistic assessments and incomplete criteria weights. To overcome incompleteness, optimal weights are derived using evidential reasoning and genetic algorithms in parallel for the construction of a non-linear programming model. Moreover, instead of crisply defined clusters, Wang considered a fuzzy clustering model for assigning weights to the preferences of experts in accordance with their membership to one or more clusters. The linguistic 2-tuple model and its aggregation operators are adopted in the author's work to conflate linguistic information.

Liu et al. considered the clustering problem in complex MCLGDM problems [83, 86, 89]. Three of their studies are reviewed below:

- In [83], considering Interval-valued Intuitionistic Fuzzy (IIF) decision matrices, an approach is presented to classify experts predicated on interest groups or risk attitude levels, in a complex MCLGDM setting. A partial binary tree DEA-DA cyclic is classifier is trained and modeled to this end. The two-stage followed by the classifier is visually illustrated in Fig. 4.5. Classification results are treated as subgroups, which are assigned weights to aggregate decision information accordingly. IIF assessments, each of which express both satisfaction and

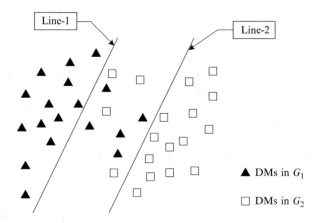

FIRST STAGE

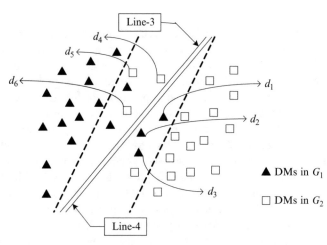

SECOND STAGE

Fig. 4.5 Two-stage partial binary tree DEA-DA classification for subgroup clustering proposed by Liu et al. (source: [83]). "DMs" is used to refer to Decision Makers (experts)

non-satisfaction degrees as pairs of fuzzy membership intervals, are accurately transformed into numerical assessments based on the risk attitude of experts, which in turn is determined through a questionnaire. The methodology is illustrated to decide on the construction of a major hydropower station.

- In [86], it is assumed that not every cluster member's individual opinion will be equally consistent with the cluster aggregated opinion, and so they are assigned distinct weights in a way that maximizes the consensus level in the cluster. Clusters are also deemed as having different importance weights than each other, due not only to their diverse sizes, but also owing to the differing usefulness in the opinion information contained in them. The individual weight of an expert is calculated based on (1) her/his weight in the cluster, computed via the minimized variance model; and (2) the weight of the cluster where (s)he belongs (calculated based on the entropy weight model, which measures the quality of the cluster information). Unlike [83], here assessments in the decision matrices are modeled by linguistic 2-tuple intervals.
- Although the above method considers the consistency level of opinions, in [89] such consistency is also preserved in the aggregation of cluster-level preferences by determining optimal cluster weights. To do this, a mathematical programming model with consistency conditions is defined. The final expert weight (which again is obtained by aggregating the associated individual weight and cluster weight) becomes thus more objective and reliable.

4.2.3 Clustering Large Groups in Emergency and Risk Situations

Preference-based clustering are utilized in conjunction with optimization by Cai et al. [19] to minimize conflicts in large-group emergency decision situations. Individual opinions, expressed linguistically, are mapped into interval numbers and then an interval similarity measure is used to divide a large, heterogeneous decision group into subgroups of like-minded participants. Interestingly, Cai et al.'s work considers multi-stage decision problems related to unconventional emergencies, in which multiple temporal stages must be considered (for example, an emergency situation in a chemical plant which lasts for 3 days after the occurrence of a fire), and the individual opinions on the available alternatives/strategies may vary from one stage to another. This multi-stage dimension is considered in the proposed model by calculating both optimal cluster weights and optimal stage weights, based on relative entropy. The resulting twofold weighting method helps mitigating conflicts and improving opinion consistency at group level. Limitations pointed out by the authors in their study include (1) not considering multiple attributes, which is a commonly present aspect in emergencies, and (2) not reflecting the distinct attitudes of decision makers.

Following their initial contributions to LGDM in general and to subgroup clustering in particular, Xu et al. in [161, 163] devoted their efforts to subgroup-based models with opinions expressed as trapezoidal fuzzy numbers in emergency LGDM:

- In [161], Xu et al. addressed a subgroup partitioning consensual decision making setting. They defined a measure of conflict between pairs of experts, given by the distance between two Interval Intuitionistic Trapezoidal Fuzzy Numbers (IITFNs), one of the many extension of Zadeh's fuzzy sets for uncertain preference modeling. Conflict at group level is likewise measured, predicated on (1) the clusters that are previously constructed based on minimizing the individual conflicts between individuals in the same cluster, as well as (2) the number of participants in them. Since a CRP takes place in their model too, once consensus has been measured and where it is not sufficient, a specific adjusted value on identified assessments is suggested to *all* members in the affected cluster. The identification of discordant preferences is heavily based on the preference of the largest cluster(s). Therefore, the overall consensus building procedure in [161] is additionally assumed to adopt the majority rule principle to some extent.

- Later on, in [163] the preference IITFNs modeling was extended into preference vectors of Generalized Interval Trapezoidal Fuzzy Numbers (GITFNs). Therefore, a GITFN similarity is proposed to identify subgroups under a fuzzy clustering setting. Particular care is taken in preserving the consistency of the aggregated collective preferences, so as to ensure effective emergency response in their target decision making scenarios. Criteria weights are unknown in the emergency MCLGDM framework considered, and the experts are assumed to have equal importance, hence clusters are weighted according to their size solely. The TOPSIS method is utilized to rank alternatives in the selection phase.

The authors also addressed in [162] emergency situations where the elimination of risk becomes of vital importance. They postulated that there exist few studies on risk coordination and mitigation in emergency decision making. The severity in the consequences of a bad decision that underestimates any component of risk is therefore highlighted. One of the major causes for risk are the heterogeneity of experts in their background, attitudes and expertise. This situation accentuates in LGDM, as stated by Xu et al., particularly with participating individuals across disciplines, regions, sectors and cultural values. On the one hand, the overall framework from [161] is enhanced by presenting an approach to manage and eliminate risk in disaster management. Risk characteristics are analyzed by combining multiple decision factors, although risk itself is measured as the level of conflict among experts (namely individual conflicts with the collective opinion), i.e. as the absence of agreement. Reduction of risk is therefore made by identifying and advising the most discordant experts across the large group, in an analogous manner as a consensus feedback mechanism. On the other hand, the clustering algorithm in [159] is adopted. The cluster majority principle is implemented to assign weights to their members.

4.2.4   Clustering Methods Under Fuzziness

Tapia-Rosero et al. developed multiple models based on subgroup clustering and opinion information modeled by fuzzy membership functions [138–140]. A common major goal among their works—described below— is to identify relevant opinions and trends across a large group, showcasing their feasibility in new product evaluation problems:

- In [139], the authors presented a method based on shape similarity to split a large group into clusters. Although experts' opinions are originally modeled as fuzzy trapezoidal membership functions, a shape-symbolic notation is defined to map each fuzzy membership function into its associated symbolic representation describing shape characteristics: slope, preference levels, and relative lengths of segments. Intuitively, individuals with similarly shaped membership functions are mapped onto similar or identical shape symbolic representations, whereby clusters describing similar opinions or trends are formed. This work concentrates only on the opinion detection process, and thus does not study the actual LGDM method with opinion aggregation.
- In [138], the above method is extended to determine the relevance of each cluster based on its cohesion, number of members and number of noticeable opinions. The cohesion measure in turn is aimed at quantifying the level of "togetherness" (i.e. agreement) between fuzzy membership functions within the cluster. Another notable contribution in [138] is the inclusion of soft computing and an LSP (Logic Scoring of Preference) aggregation method to combine individual membership functions and obtain evaluation information (e.g. on a new product proposal) at cluster level. Based on generalized conjunction and disjunction, LSP can reflect aspects of human decision making.
- Afterwards, Tapia et al. presented in [140] a novel framework that, besides integrating the characteristics of their above works, takes into account the uncertainty associated to the preferential information in the process of identifying relevant opinions. A more meaningful selection of relevant opinions or clusters of them (e.g. trends) is thus made on the basis of analyzing not only the actual evaluations given to a new product proposal, but also taking account of their inherent level of uncertainty.

The doctoral thesis elaborated by Tapia-Rosero [141], entitled "Fusion of preferences from different perspectives in a decision-making context" compiles the three above studies. The thesis investigates the problem of handling complexity in decision making under a large volume of preferential information and considering multiple perspectives. Three key research questions are postulated: (1) how to deal with a large number of preferences, (2) how to identify and evaluate relevant opinions, and (3) how to combine opinions stemming from multiple perspectives. Questions (1) and (2) are addressed in [138–140]. A fusion method that takes into account different backgrounds, e.g. educational levels, expertise areas and personal profiles, is defined to address the third question. The fusion approach introduces the concept of Decision Making Unit, laid out as a hierarchical structure that enables the efficient propagation of preference information.

Another fuzzy clustering approach, in particular based on the Fuzzy Equivalence Clustering (FEC) method, was presented in [152] by Wu and Liu. The FCM algorithms are suitable to handle fuzzy linguistic information, which in [152] is gathered in the form of interval type-2 fuzzy decision matrices. It also produces dynamic clustering results efficiently, thereby overcoming the need for specifying a number of clusters ahead of the clustering process (contrary to e.g. FCM). The aggregation strategy to obtain each cluster preference relies on the Combined Weighted Geometric Averaging (CWGA) operator extended to interval type-two assessments, and it takes into account both the importance of individuals and the relative positions of their opinions in the cluster. Furthermore, a ranking method of interval-valued type-two fuzzy set assessments is defined to obtain easily comparable values across alternatives in the large group. Interestingly, this is in addition one of the few approaches that consider individually assigned criteria weights, i.e. different participants may have different perspectives about the relative importance of the evaluation criteria [112]. This is proved to be the case in their target application example on senior employee hiring, where for instance a member of the recruitment panel can consider the candidate's communication skills as more important, whilst another could deem her/his computer skills as more relevant.

Further research has been recently conducted by Wu et al. on clustering interval type-2 fuzzy preferences [153, 154], with diverse applications:

- In [153], they coined a new decision framework so-called *Double large-scale GDM*: a framework involving both a large group of experts and a large number of evaluation criteria. A linguistic extension of the PCA (Principal Component Analysis) method for data dimensionality reduction, is proposed to determine a reasonable number of relevant criteria predicated on possible inter-dependences between the original criteria. The FEC method is then applied as part of the process of aggregating the preferences of (many) experts. Questionnaire surveys are used to obtain a codebox, i.e. an information structure that allows participants to express their opinions freely, using natural language, and mapping such opinions into interval type-2 fuzzy set decision matrices. This codebox and a sample matrix (also obtained via questionnaires) are the necessary inputs to apply the linguistic PCA for criteria reduction. The study is applied to a crowd-based problem of customer decision making in e-commerce. Furthermore, the linguistic CWGA operator is utilized for aggregation.
- In [154] another remarkable aspect widely considered in subsequent LGDM developments, is introduced: the use of Social Network Analysis (SNA). Methods for SNA are reported as useful for addressing LGDM problems despite, as Wu et al. argued, most related LGDM studies assumed that experts are socially independent from each other. The Louvain method [12], popularly known owing to its effectiveness and ease of implementation for community detection in large

networks, is adopted for the task of identifying clusters. The Louvain method is composed by two steps: a network partition step and a partition combination step, which in Wu et al.'s framework facilitate the simplification of the large group structure, thereby reducing its complexity—in terms of social connections and opinions—and dimensionality. Weights of experts are calculated based on their reputation in the social network structure. Another contribution is the definition of a new interval type-2 fuzzy TOPSIS model for LGDM in a complex and uncertain setting, that aggregates preferences via the interval type-2 fuzzy WAM operator. An example of application is provided for the problem of choosing the best restaurant to celebrate a business dinner.

4.2.5 Other Notable Contributions to Subgroup Clustering in LGDM

Xu et al. introduced in [159] an entropy-driven weighting method to pre-assess preferential information provided by experts, and then classifying experts into several clusters. To alleviate the cost of making a large group decision, members who do not provide enough relevant information to the decision problem are removed before the preference aggregation process. The remaining participants are weighted according to the size of the cluster they belong to. The principle followed for this, similarly to other reviewed works in this section, is that a larger cluster means a higher weight being assigned to its members. Numerical preference vectors in the unit interval are considered, along with the weighted averaging operator to aggregate them. The method is illustrated in a decision problem on selecting the best investment alternative in a large company.

Failure Mode and Effect Analysis (FMEA) [136] has been recently investigated from a clustering LGDM perspective. FMEA is a qualitative analytic technique characterized by multi-expert brainstorming and elicitation of comments (usually in a tabular format), such that experts identify possible causes for failure in a product, the likelihood of occurrence for such failures, their impact and how easy they would be to detect, among other factors. Liu et al. argued in [80] that classical FMEA focuses solely on small groups of experts. However in nowadays organizations composed by multiple departments, a larger number of experts are often required. Furthermore, existing FMEA techniques exhibit limitations in assessing and ranking failure modes, as well as weighting risk factors (e.g. via the entropy method). To overcome these issues, the authors presented in [80] a threefold contribution: (1) a similarity-based clustering method based on hesitant fuzzy linguistic preferential information, where assessments may show hesitancy between two or more linguistic terms, (2) a within-cluster consensus measure to assign cluster weights according with the level of consensus among cluster members, and (3) a novel risk prioritization approach using cluster analysis. Their framework is applied to quality and reliability planning FMEA in the health care risk assessment domain.

Liu at al. [87] assumed in their work the existence a priori of multiple groups of participants forming a large decision group. Nevertheless, their approach relies in cluster analysis as the driving factor to assign weights to these multiple groups. Once individuals provide their preference vectors on a discrete numerical scale (associated to a linguistic scale), an aggregation process based on counting is conducted to obtain, for each (sub)group, a percentage distribution of its members' evaluations for each alternative. The percentage distribution is analyzed to determine the consensus level within each group, which in turn is used to yield the objective importance weight of that group. A subjective group weight is also assigned and provided by the organizer of the large group decision problem, thereby combining both the objective and subjective weight of each group into a comprehensive decision weight. Finally, a weighted averaging aggregation is made to aggregate each group's percentage distribution into the large-group percentage distribution, which is exploited predicated on the Promethee II selection method to obtain pairwise comparisons between alternatives and make a final decision. The proposed method for participators from multiple groups is utilized in a case study involving local energy network dispatch decisions.

The work by Zhu et al. [194] constitutes an interesting approach for GDM at a very large scale, where numerous subgroups may exist and the most sensible solution relies in finding a single cluster representative preference for each subgroup (even though some individual opinion information is partly lost as a result). This facilitates the process of aggregating information in terms of cost and complexity, two aspects requiring attention when the volume of relevant decision information becomes large. It is noteworthy that the study focuses on the processes of clustering the large group and simplifying it by leaving out non representative information, hence the aggregation, exploitation or consensus aspects are not investigated. Two dimensions of information are considered: preferential (decision matrices considering multiple evaluation criteria) and referential (preference relations judging pairs of alternatives). Three dimensional gray correlation indexes are then defined to comprehensively measure the similarity between pairs of participants and cluster the initial large group based on the mentioned "double" information, alongside a hierarchical clustering procedure. A strategic decision by a large firm is the target scenario shown by Zhu et al. in their study.

Soto et al. [134] proposed a multiple-group decision making method for heterogeneous groups of participants (affinity groups) forming a structured social network, and considering multiple rounds of discussion on preferences. Social relationships are analyzed to discover affinity groups, deemed as (sub)groups of experts. In addition, Data Science and Sentiment Analysis techniques play an important role in Soto et al.'s method, for constructing profiles based on reputation and confidence information. Profiles are in turn used for weighting experts' preferences. Notably, this work faces the problem of meaningfully discriminating between individual opinions from participants with different expertise in a social network setting. Although preferences are modeled qualitatively and encoded into linguistic values in

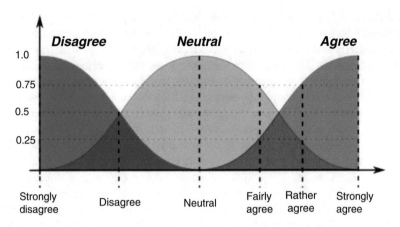

Fig. 4.6 Fuzzy sets for encoding responses to a Likert-scale question (source: [134])

an agree-disagree Likert scale (defined by logistic curves, see Fig. 4.6), the method can be easily extended to handle numerical opinions. Likewise, a fuzzy aggregation algorithm based on membership to affinity groups, previous participation history and expertise, is yielded, but an overview of relevant linguistic aggregation operators is also provided. Their method was implemented into a platform called "moviQuest-MAS".

Finally, in [168] Xu et al. presented a novel two-stage approach for consensus-based MCLGDM, with the following characteristics: (1) the large group is divided into subgroups based on Self Organizing Maps (SOMs) [74], (2) weights yielded by the SOM clustering approach are used to aggregate individual preferences into cluster preferences, (3) each cluster preference is treated as an individual preference, hence the LGDM problem is "converted" into a small-group problem. Before executing the SOM algorithm, preferences initially expressed as numerical decision matrices are transformed into vectors, aggregating assessments over criteria (hence not all the original opinion information is fully considered in the SOM algorithm). Xu et al.'s proposed CRP is aimed at reducing the number of consensus rounds required, which is an important aspect to consider in emergency situations such as earthquake shelter selection.

4.3   LGDM Methods

Surveyed works on LGDM methods are summarized in Table 4.3 and subdivided in the following six themes:

- Methods for complex MCLGDM.
- Aggregations based on mutual assessment support in LGDM.
- LGDM methods with fuzzy membership-based opinions.

Table 4.3 Summary of surveyed works on LGDM methods

Theme	Representative publications (listed by theme alphabetically)
Methods for complex MCLGDM	Liu et al. [81, 82, 84, 85]
Aggregations based on mutual assessment support in LGDM	Zhang et al. [185]
LGDM methods with fuzzy membership-based opinions	Tapia-Rosero et al. [142]
Estimating incomplete assessment and weight information in LGDM	Li et al. [78]; Xu et al. [166, 167]
LGDM with Linguistic Distribution Assessments	Yu et al. [175]; Zhang et al. [189]
LGDM with Double Hierarchy Hesitant Fuzzy Linguistic Information	Gou et al. [52]

- Estimating incomplete assessment and weight information in LGDM.
- LGDM with Linguistic Distribution Assessments.
- LGDM with Double Hierarchy Hesitant Fuzzy Linguistic Information.

4.3.1 Methods for Complex MCLGDM

Liu et al.'s research includes numerous contributions for improving complex MCLGDM methods. Some of their works are summarized below [81, 82, 84, 85]:

- In [81], they firstly introduced their complex MCLGDM setting as a decision framework in which more than 20 participants situated at different places (and possibly distinct times) need to make a joint decision in a problem where connections exist among criteria. The complexity in their framework is threefold: complexity in the decision attributes, complexity in the profile of participants, and complexity in the decision environment. The authors' contribution focuses on the aggregation and simplification of an initially large set of criteria weights: a Partial Least Squares (PLS) method is defined to aggregate a considerable number of criteria into new ones deemed as *latent variables* for decision making. The weights of the reduced set of representative criteria are also determined. Experts used Interval-valued Intuitionistic Fuzzy (IIF) assessments to provide their preferences in decision matrices. With the help of an improved $C - OWA$ operator, the information expressed as IIF assessments is transformed into single numerical assessments with minimum information loss, as a requirement for using the PLS algorithm. Similarly to their previously discussed work in [85], the selected real-world problem to validate their method concerns the choice of the best project proposal on building a major hydro-power station.
- In [82], Liu et al. defined an extension of their above method. In particular, a two-stage PLS path modeling was formulated, with the added value of eliminating correlations among indicators stemming from the subjectivity and heterogeneity

of the decision group. Both direct and inverse correlations are overcome in their extended approach. A simplifying assumption is made on the modeling of preferences, which are straightforwardly expressed as numerical utility vectors, whereby PLS can be applied more directly.

- The study presented in [84] states that complex MCLGDM can be applied at three levels: (1) individual level, e.g. taking every citizen's information into consideration; (2) company level, e.g. in economic industrial sectors with many companies involved; and (3) state level, e.g. in international organizations such as the United Nations[1] (UN). In this case, an enhanced version of Principal Component Analysis to deal with IIF information, called IIF-PCA, is introduced as an alternative approach to identify relevant independent decision criteria. The experts and original criteria have all equal importance, but after applying IIF-PCA relevant components may have distinct weights depending on their level of contribution to discriminating among the available alternatives (this information is extracted from the eigenvalues associated to the evaluation criteria in PCA).

- Lastly, an extension of [84] was provided in [85], by incorporating an additional case study on major hydro-power station selection, and enabling a direct preference aggregation strategy on IIFs, by using the IIFWA and IIFWG operators (weighted extensions of the arithmetic and geometric means for aggregating IIFs).

4.3.2 Aggregations Based on Mutual Assessment Support in LGDM

Zhang et al. [185] investigated various types of aggregation models for GDM and LGDM with ordinal information. Most remarkably, however, they faced the following challenge: most aggregation operators ignore the degree of support between pairs of experts' assessments during the aggregation process into a collective assessment. The authors argued that managing conflicting opinions in ordinal problems at large scale may sometimes make it difficult to reach consensus due to the presence of non-cooperative behaviors. In order to overcome this situation, a two-stage dynamic method for decision making under ordinal information is presented, featuring a data cleansing motivated by the concept of soft consensus [67] to eliminate conflicts among ordinal preferences (converted into fuzzy preference relations). An additional contribution to their core LGDM method is an automatic approach to build consensus based on the Power Averaging (PA) aggregation operator. The PA operator enables aggregation inputs to support and reinforce each other [8], which is an important aspect to consider when similar opinions to each other should be emphasized. For the selection process, a new non-parameterized support function based on rough sets dominance is introduced for LGDM problems

---

[1]United Nations website: http://www.un.org/en/index.html.

with ordinal preferences, and its performance is demonstrated on an environmental decision making problem related to control degradation in a river basis. Further research from the authors on behavior management is referenced in Sect. 4.5.

4.3.3 LGDM Methods with Fuzzy Membership-Based Opinions

Tapia-Rosero et al. concentrated most of their LGDM research on developing fuzzy membership-based clustering methods, as previously explained in Sect. 4.2. Notwithstanding, they also developed an innovative preference aggregation method [142] that assumes opinions expressed as trapezoidal fuzzy membership functions, whose shape characteristics (see Fig. 4.7) are encoded into a symbolic linguistic representation. In their large group setting, the opinions can be supplied by participants with diverse levels of knowledge (students, non-experts and professionals), areas of expertise (such as engineering, marketing or design), or personal backgrounds e.g. different family and marital status. The process of gathering opinions across participants might also vary, e.g. surveys, polling, social network posts, etc. The study identifies an important gap in the domain of organizational decision making, namely on how to fuse the preferences provided by a body of customers (product consumers) and the preferential constraints adopted by managerial staff (product providers). The gap is bridged by proposing a LGDM method that fuses preferences from large groups of potential customers and those from corporate managers into a multi-level setting, whilst reflecting the different individual points of view—based on knowledge and expertise areas—as faithfully as possible. This is typically the case of multi-national companies with nationwide and regional headquarters for instance. The solution for this multi-level and multi-perspective scenario is a recursive aggregation methodology based on Logic Scoring of Preference (LSP) method, which is in turn founded on the Generalized Conjunction/Disjunction (GCD) aggregation operators. This facilitates the division of an initially complex and large-scale problem into more manageable subproblems. The underlying aggregation operators can reflect the relative importance of evaluation criteria. Further,

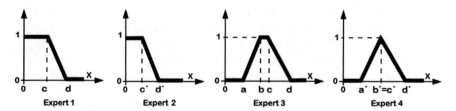

Fig. 4.7 Example of experts' opinions expressed by means of trapezoidal, semi-trapezoidal and triangular fuzzy membership functions defined by parameters a, b, c, d, whose shape characteristics are studied by Tapia-Rosero et al. (source: [142])

the resulting model can efficiently incorporate the "wisdom of the crowd" in terms of opinions from a large body of social media customers, whilst the decision power of managerial participants is naturally kept.

4.3.4 Estimating Incomplete Assessment and Weight Information in LGDM

The following two contributions by Xu et al. [166, 167] consider the problem of estimating missing decision information, and they are integrated with previously existing approaches for subgroup clustering. We refer the reader to Sect. 4.2 for a discussion of this method along with other ones:

- The ideas from clustering methods in previous works [161–163] are adopted by Xu et al. to propose in [166] a method to estimate missing information in incomplete decision matrices, in scenarios where there exist social trust information among experts, hence direct and indirect trust values can be derived and thus the missing assessments can be estimated. Moreover, to improve and make the clustering results more reasonable, the distance similarity formula is redefined based on the analysis of a previous decision problems history, which combined with a cosine similarity led the authors to propose a double clustering model.
- Xu et al. [167] also extended the two-layer weight determination method for experts' weights based on subgroup clustering and initially presented in [86]. Considering preferences elicited as interval 2-tuple linguistic decision matrices, their extended method focuses on utilizing the fuzzy entropy and relative entropy measures to establish cluster weights and also to exploit collective linguistic information for obtaining the solution alternative(s).

Li et al. [78] also studied the problem of calculating missing criteria and expert weights information for MCLGDM. In their approach, the computation of criteria weights is founded upon entropy measures from information theory [129]. As for experts' weights—deemed as difficult to obtain in a large group—they are determined by a non-linear programming optimization model. Both weighting approaches are set into a TOPSIS framework that considers closeness degrees to a Positive Ideal Solution (PIS) and Negative Ideal Solution (NIS). The divergence between Interval-valued Intuitionistic Fuzzy assessments and the PIS and NIS are quantified by fuzzy cross entropy functions. An application example is presented for selecting the best customer-oriented service in automobile manufacturing.

4.3.5 LGDM with Linguistic Distribution Assessments

A Linguistic Distribution Assessment [186] is defined upon a linguistic term set $S = \{s_0, \ldots, s_g\}$ (see Chap. 2, Sect. 2.3) as an assessment $p = \{\langle s_h, \beta_h \rangle | s_h \in S\}$, such that $\beta_h \in [0, 1]$ indicates the degree to which the assessment complies with the linguistic term s_h, and $\sum_h \beta_h = 1$.

Yu et al. revisited in [175] the concept of unbalanced Linguistic LDA for their use in LGDM. Unbalanced LDAs were originally introduced in [34] as an extension of the original LDAs [186] that can reflect aggregated collective information from linguistic assessments expressed using multigranular unbalanced linguistic term sets. Multigranular unbalanced linguistic information can take distinct psychological perceptions of decision makers into account. Owing to the unbalanced LDA model, the LGDM method by Yu et al. does not require to unify unbalanced multigranular linguistic assessments coming from diverse experts, and all the original individual information is preserved as much as possible. An extension of the TODIM[2] method [49] that incorporates unbalanced LDAs is proposed, integrating an iterative algorithm to construct unbalanced LDA decision matrices upon individual opinions. Further, gain and loss measures are defined on unbalanced LDAs, whereby the best alternative is selected.

Simultaneous advances on LGDM methodologies with LDAs a have been made by Zhang et al. in [189], in a large group context characterized by participants with different levels of knowledge, who require using linguistic term sets with different levels of granularity. Specifically, the work presents a new linguistic computational model for MCLGDM that handles multigranular linguistic preferences by keeping a maximum level of information, using the LDA model and its associated aggregation operators [186]. The results of the LDA-driven selection process provide interpretable and accurate results, in compliance with the well-known linguistic 2-tuple model. The fusion process involves two pre-aggregation steps: (1) transforming individual linguistic 2-tuple assessments into LDAs and then unifying multigranular LDAs into a common linguistic term set without loss of information. The individual decision matrices containing unified LDA assessments are subsequently aggregated via the DAWA (Distribution Assessment Weighted Averaging) operator. The original distance measure between two LDAs is likewise reformulated and improved. The study includes a remarkable theoretical comparison against similar MCLGDM methods and linguistic information modeling approaches.

4.3.6   LGDM with Double Hierarchy Hesitant Fuzzy Linguistic Information

Double Hierarchy Hesitant Fuzzy Linguistic preferences have been recently introduced by Gou et al. in [52] to provide an answer to the question: how to express linguistic preferential information—such as "only a little low" or "far from very high" for instance—in a more precise and intuitive manner? Their approach consists in modeling the linguistic information upon a twofold hierarchy called Double Hierarchy Linguistic Term Set (DHLTS), composed by:

---

[2]The name TODIM originates from the Portuguese acronym for Interactive and Multiple Attribute Decision Making.

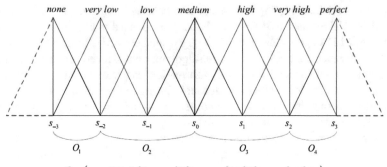

$$O_1 = \{o_0 = just\ right, o_1 = a\ little, o_2 = only\ a\ little, o_3 = far\ from\}$$

$$O_2 = \{o_{-3} = entirely, o_{-2} = very\ much, o_{-1} = much, o_0 = just\ right, o_1 = a\ little, o_2 = only\ a\ little, o_3 = far\ from\}$$

$$O_3 = \{o_{-3} = far\ from, o_{-2} = only\ a\ little, o_{-1} = a\ little, o_0 = just\ right, o_1 = much, o_2 = very\ much, o_3 = entirely\}$$

$$O_4 = \{o_{-3} = far\ from, o_{-2} = only\ a\ little, o_{-1} = a\ little, o_0 = just\ right\}$$

Fig. 4.8 Modeling of double hierarchy linguistic information (source: [52])

1. A basic linguistic term hierarchy, e.g. "very low", "low", ... "very high".
2. An auxiliary linguistic term hierarchy, including expressions to facilitate a richer expressiveness when combined with terms in the basic hierarchy, e.g. "only a little", "far from", etc.

Figure 4.8 illustrates the modeling process of double hierarchy linguistic information. The DHLTS allows for more intuitive and accurate linguistic information, but it is also extended in [52] to accommodate situations of hesitancy, where an expert may hesitate between terms in the basic hierarchy or between expressions in the auxiliary hierarchy to judge the same term in the basic hierarchy. Accordingly, Gou et al. present a framework for consensual LGDM with double hierarchy hesitant fuzzy linguistic preference relations, that considers a prior clustering of the large group based on an entropy measure, and then incorporates a consensus model with an underlying consensus measure among double hierarchy hesitant assessments. The complete model is illustrated in an water resources management application in Sichuan province (China).

4.4   Consensus in LGDM

Surveyed works on consensus for LGDM are summarized in Table 4.4 and subdivided in the following eight themes:

- Semi-supervised consensus support approaches.
- Consensus in emergency LGDM.
- Consensus building under social data and opinion dynamics.

Table 4.4 Summary of surveyed works on consensus in LGDM

Theme	Representative publications (listed by theme alphabetically)
Semi-supervised consensus support approaches	Palomares et al. [102, 104]
Consensus in emergency LGDM	Xiang [158]; Xu et al. [164]
	Xu et al. [164]
Consensus building under social data and opinion dynamics	Dong et al. [37, 39];
	Gupta [53];
	Wu et al. [151]
Consensus for 2-rank LGDM problems	Zhang et al. [190]
Consensus on individual concerns and satisfactions	Zhang et al. [191]
Consensus and consistency under linguistic information and anonymity preservation	Zhang [188]
Consensus with changeable subgroups of participants	Wu and Xu [155]
Exploring classical consensus models in LGDM	Labella et al. [75]

- Consensus for 2-rank LGDM problems.
- Consensus on individual concerns and satisfactions.
- Consensus and consistency under linguistic information and anonymity preservation.
- Consensus with changeable subgroups of participants.
- Exploring classical consensus models in LGDM.

4.4.1 Semi-supervised Consensus Support Approaches

In spite of the sheer multitude of consensus support approaches developed over the last two decades, the applicability for most of them was demonstrated almost exclusively to solve problems involving small groups. Indeed, it was not until recently that the first consensus models and CSS envisaged specifically for supporting large group decision groups were proposed. The works by Palomares et al. [102, 104] reflect the initial efforts in this trend.

- A multi-agent architecture to support cost-efficient CRPs in LGDM was presented in [104]. The architecture combines a standard consensus model based on a feedback mechanism, with a set of autonomous agents capable of substituting the human moderator and conducting the major supervisory steps of the CRP by themselves. However, the most distinctive feature in Palomares et al.'s architecture is twofold.
  Firstly, the so-called "expert agents" are integrated as autonomous entities each of which are capable of acting on behalf of the human participant they represent. Each expert agent adopts a behavioral profile that reflects the expert behavior in modifying assessments over the course of the CRP, out of three different

Fig. 4.9 Example of change
functions for (**a**) sure profile,
(**b**) unsure profile, and (**c**)
neutral profile (source: [104])

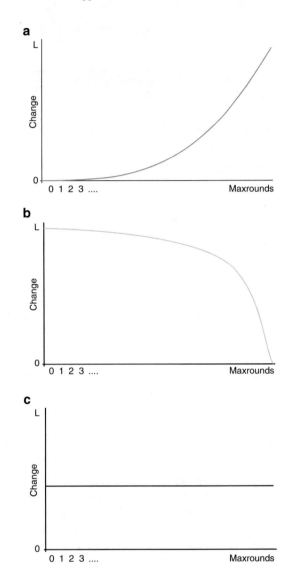

profile types. A change profile is mathematically defined by a simple one-variable
negotiation function called *change function*, inspired by *Kasbah* [26], describing
how prone an expert is to adjust her/his opinions at each round of the CRP (see
Fig. 4.9).

Secondly, supervision $IF - THEN$ rules are defined to decide, for each
advice of feedback on a single expert assessment p_i^{lk}, whether (1) the human
expert should revise and accept/decline the feedback herself, or instead (2)
delegating into her associated expert agent the task of adjusting the assessment
p_i^{lk}. In particular, where the advice suggested implies a drastic change in the

pairwise preference between the two alternatives x_l and x_k (recall that additive preference relations are used), then human supervision is requested. Otherwise the expert agent autonomously accepts and applies the change on p_i^{lk}. Intuitively, without the supervision rules the expert agents have the potential to completely replace the human expert in the task of revising preferences, thereby turning the consensus model into an approach based on automatic updates of preferences. Notwithstanding, Palomares et al. consider that in many real-world LGDM scenarios the sovereignty in terms of the control of participants on their own opinions must be preserved to some degree.

The proposed solution based on combining autonomous agent profiles with supervision rules ensures a trade-off between maintaining expert participation and fostering cost-effective consensus reaching. The experimental results showed a dramatic decrease in the temporal cost invested, average number of assessment supervisions required per expert, and average number of experts requiring the supervision on (some of) their preferences per CRP round.

- A desktop implementation of the above semi-supervised multi-agent architecture incorporating a Web service interface, was showcased in [102] and utilized by a student cohort to consensually select the best destination for their planned final year trip.

4.4.2 Consensus in Emergency LGDM

Shortly after the above described early works on consensus for LGDM, in [164] Xu et al. introduced an exit-delegation mechanism into a dynamical consensus model that handles clusters of participants under different levels of emergency decision making. Once a preference-based clustering process is conducted, consistency and consensus are measured as the two leading criteria for building agreement. Within the scope of this work, consistency means that there is no contradictory information in the fuzzy preference relations associated to individuals, and it is analyzed to prevent inconsistent group preferences after aggregating. Experts and clusters are therefore weighted predicated on the consistency underlying their preferences. The main contribution however is the exit-delegation mechanism, which—under some specific consensus conditions and time constraints—may invite some participants from a specific cluster to abandon the LGDM session. The innovative delegation mechanism can accordingly be used by these clusters to give degrees of trust (weights) to other clusters of experts who remain in the decision process, thereby preserving their concerns to some degrees until the final decision is made.

Xiang in [158] considered the emergency situation of local energy shortage after a disaster, a decision situation characterized by time limitations. Existing decision support methods in this domain fail to allow and handle large group consensual decisions, as the processes may become time consuming, expensive and impracticable. In accordance with this, Xiang presented an energy network dispatch optimization model with a Web tool for automatic LGDM under rational consensus.

Once the initial preferences (numerical decision matrices in a discrete scale) are submitted, a consensus response plan is obtained within limited time. Prior to this, the optimization method is executed to determine a suitable set of alternatives for emergency energy shortage strategy. There is an a priori assumed partition of the large group into subgroups with associated weights. The automatic consensus building module assumes total cooperativeness for applying rational automatic adjustments on assessments. Interestingly, the consensus measure relies on distances to collective information at three levels: (1) distances between individuals and subgroups, (2) distances between subgroups and the large group, and (3) distances between individuals and the large group.

4.4.3 Consensus Building Under Social Data and Opinion Dynamics

Simultaneously with the above described developments, new consensus models for LGDM arose which introduced the existence of social trust relationship information. Wu et al. [151] proposed a trust-driven consensus model for decision making in social networks under a 2-tuple linguistic context. Due to the uncertainty and partial knowledge about the decision problem at hand, some experts in the social network might be unable to assess some alternatives under certain evaluation criteria, thus providing incomplete linguistic decision matrices. The authors defined a social trust propagation method to derive the unknown individual assessments via trusted third parties. A *<trust,distrust>* pair of values is utilized to determine the trust score and knowledge deficit (level of ignorance) associated to each individual. Incomplete values are then estimated, predicated on other trusted experts' knowledge. The consensus model advices a target modified value for each assessment requiring adjustment. Moreover, a visual feedback approach provides experts with an insightful representation of their consensus status in the group, together with feedback to modify their opinions. Simulations experiments and a case study for green energy supplier evaluation, demonstrate the visualization of the predicted consensus building status in following CRP rounds.

Remark 4.2 Different GDM, LGDM and consensus-based methods have been proposed to deal with incomplete preferential information. For a comprehensive overview of such approaches, we refer the interested reader to the survey presented by Ureña et al. [147].

Taking another step forward in the consideration of social relationship information, the study carried out by Dong et al. in [37] bridges the gap between opinion dynamics and consensus for LGDM. Dong et al. devised a consensus building model suitable for an opinion dynamics setting whereby individuals are connected to each other under a network structure modeled as a directed graph, and they update their individual opinions predicated on the opinions shown by other trusted individuals connected with them. The study attempts to answer three questions:

1. Who are the leaders, i.e. the persons whose influence shall determine the consensus opinion?
2. How to construct a social network structure that ensures the formation of consensus?
3. How to provide rules to move individual opinions in a network to form consensus, with the involvement of the moderators?

To answer the above questions, and assuming that every individual (agent) trusts at least another individual, in [37] an opinion dynamics model is presented to build consensus by constructing a network capable of holding a number of required consensus conditions. Two steps are undertaken: network partition, and addition of new edges to the partitioned network. Inspired by traditional opinion dynamic models, opinions are expressed as single numerical values in a continuous domain, hence the consensus opinion would consist in a value in the defined interval. The process of aggregating preferences boils down to fusion rules taken from well-known opinion dynamic models, e.g. the deGroot model.

Dong et al. also presented in [39] a review of consensus processes applicable to both GDM and LGDM situations taking place under social network settings. They classify existing approaches for social network-based GDM (abbreviated as SNGDM in their work) into two paradigms:

(a) *Consensus building approaches based on trust relationships* [130, 192]. These approaches integrate information about trust levels between (pairs of) individuals in the CRP and its underlying consensus measurement and consensus building stages. Normally, such trust information is used in the process of weighting experts' preferences prior to their aggregation. Aspects of study include the definition of trust propagation operators [130], missing trust value estimation methods and trust data fusion processes [192], aimed at obtaining complete trust information across the social network. Figure 4.10a shows a general framework for defining consensus models founded on trust relationship information.

(b) *Consensus building approaches based on opinion evolution* [37]. Opinion evolution dates back to French and John's pioneering model in 1956 [45], having evolved towards nowadays' opinion dynamics research. Models to conduct CRPs under opinion evolution principles have still been very scarcely investigated as of yet. Their main innovation relies in the modification of individual opinions as a result of numerous small interactions among participants (usually pairs of them), as opposed to most extant consensus models in which this opinion modification process is controlled by either a CRP moderator or an implemented consensus model acting as such. The stabilization of opinions into consensus, polarization or fragmentation situations, as well as the management of uncertainty and vagueness in evolving opinions [33], are a key aspect of research in this setting. A general framework for CRPs based on opinion evolution process is shown in Fig. 4.10b.

Further, the survey highlights the main research challenges and research directions to adopt in social network-based decision making.

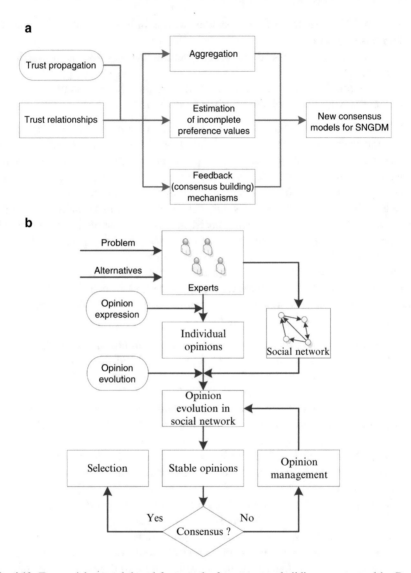

Fig. 4.10 Two social network-based frameworks for consensus building, as surveyed by Dong et al. [39]: (**a**) based on trust relationships, (**b**) based on opinion evolution

An alternative approach for consensus support in an opinion dynamics-based framework was established by Gupta in [53]. Existing consensus building methodologies are not adaptive enough to the realistic behaviors and evolution of collective consensual opinions across successive rounds of interaction. Accordingly, an iterative algorithm that manages the evolution of consensus building based on interrelationships between participants is proposed. Shifts in individual opinions are produced depending on the level of adoption of other views across the group. A graph describing the influence relationships and a fuzzy rule inference system are constructed as supporting tools to guide consensus. The fuzzy inference system acts as an automatic approach for consensus building based on adopting other members' opinions (given by linguistic preference relations). A convex combination operator is used to combine the current individual view with the group's view.

4.4.4 Consensus for 2-Rank LGDM Problems

The ultimate goal in most consensus models in LGDM (and roughly any GDM or MCDM process in general) is either to select the best alternative from a finite set of them, or to rank the alternatives from the most to the least suitable one. However, in some situations the required solution is neither a best alternative nor a complete ranking of them, but instead to make a partitioning of the solution space into suitable alternatives and unsuitable ones. Zhang et al. addressed this interesting research question in [190] under a different and novel perspective. The authors consider the 2-rank selection process for MCGDM problems, in which a two rank level solution is yielded, i.e. ranking one subset of alternatives over another. This type of solution is required in multiple real-life LGDM problems such as: (1) selecting R&D projects to be granted out of a large number of submissions, (2) rewarding a number of excellent employees for their meritorious contributions to their company over the last year, or (3) selecting five conference papers to give them the best student paper award. The model in [190] is the first effort to apply a 2-rank MCGDM process alongside a CRP: it not only considers a 2-rank selection process, but also introduces a pioneering 2-rank consensus model consisting of a 2-rank consensus measure and a 2-rank minimum adjustment rule for increasing consensus. The minimum adjustment rule is transformed into a 0–1 mixed linear programming model, which is proved to require less adjustments in practice compared to other approaches, hence the goal of more cost effective CRPs in LGDM is also investigated. Preferences are expressed linguistically, by means of multigranular linguistic decision matrices, and assessments are aggregated across individuals (for obtaining the collective opinion) and criteria (for the 2-rank exploitation phase), by using the 2-tuple WAM and OWA operators, respectively. A scheme of the 2-rank consensus model proposed by Zhang et al. is presented in Fig. 4.11.

4.4.5 Consensus on Individual Concerns and Satisfactions

Zhang et al. continued their studies on consensual LGDM, recently presenting in [191] a consensus model taking account of the concerns and satisfactions of participants from a large group. Their model admits heterogeneous preferential information expressed as preference orderings, preference vectors, fuzzy prefer-

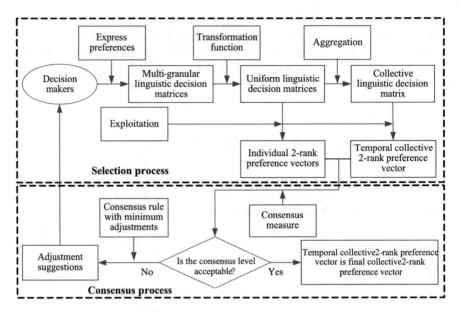

Fig. 4.11 2-Rank consensus model proposed by Zhang et al. (source: [190])

ence relations or multiplicative preference relations. Before the CRP commences, individual selection methods associated with different preference formats are conducted on heterogeneous preferences to derive preference vectors from them. The preference vectors are then used to cluster the large group into subgroups, in which the individual concerns among decision makers are considered. A consensus measurement method that considers the individual concerns on alternatives is defined for measuring the consensus degree, and the linguistic 2-tuple approach is incorporated to assess individual and collective satisfactions with the consensus degree achieved, under the premise that different experts might have distinct perceptions about the level of agreement achieved so far. As for concerns, they describe thoughts on whether some specific alternatives should be eligible for selection of not, i.e. they constitute additional constraints to ensure genuinely accepted decisions. An example is presented on using funds to improve the running conditions of a university school. The resolution framework is presented in Fig. 4.12.

4.4.6   Consensus and Consistency Under Linguistic Information and Anonymity Preservation

Consistency in LGDM was investigated in parallel with consensus reaching by Zhang in [188]. Due to the frequently large diversity of participants in LGDM, both consistency and consensus degrees are taken into account in their consensus

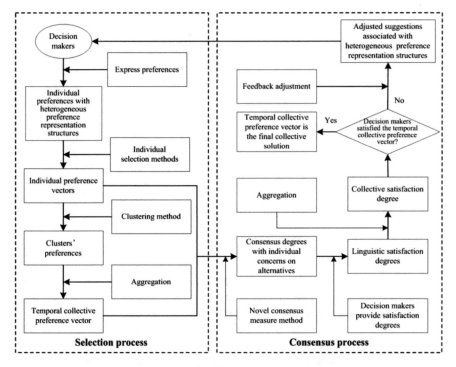

Fig. 4.12 Resolution framework of the heterogeneous consensus model for LGDM with individual concerns and satisfactions proposed by Zhang et al. (source: [191])

model, to objectively determine weights of multiple groups assumed a priori in the target framework. In order to enable the preservation of anonymity, either equal or unequal weights can be considered for experts. The defined consensus measure, which calculates pairwise similarities between experts at each group separately, can be used to determine objective weights for them and obtain the most representative large-group preference. The modeling and aggregation of preferential information is another key aspect in [188], with individual preferences expressed as linguistic preference vectors, whereas collective (aggregated) information is represented by Probabilistic Linguistic Term Sets (PLTS). Regarding the aggregation of linguistic preferences, a PL-WAA operator (Probabilistic Linguistic Weighted Arithmetic Averaging) is introduced.

4.4.7 Consensus with Changeable Subgroups of Participants

While most contemporary LGDM approaches consider static subgroups, Wu and Xu [155] studied the problem of large group consensus building under changeable clusters. Given that individual opinions tend to change over the course of a CRP, it

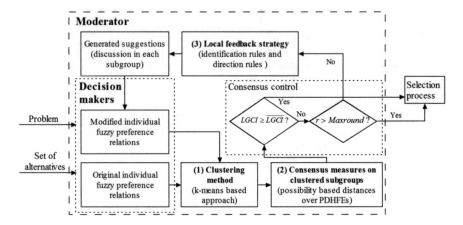

Fig. 4.13 Consensus framework proposed by Wu and Xu (source: [155])

is natural to assume that, where no explicit distinction among subgroups exists, the composition of such subgroups may sometimes change before consensus is achieved. Wu and Xu presented a novel distance measure between Possibility Distribution Hesitant Fuzzy Elements (PDHFE), which are used to accurately represent collective opinion information at subgroup level (obtained by aggregating individual fuzzy preference relations). The PDHFE distance function serves as the basis for defining a consensus measure based on distances to the collective opinion and a feedback mechanism at alternative, cluster and individual level. Identification and direction rules are defined for the changeable clusters, and the advice is generated in the form of discrete fuzzy assessments—rather than PDHFEs—to facilitate their understanding. The feedback approach also relies in the identification of ideal and anti-ideal clusters. The target application is emergency LGDM, regarding the choice of a rescue plan. Figure 4.13 outlines the consensus framework proposed by Wu and Xu in [155].

4.4.8   Exploring Classical Consensus Models in LGDM

A comparison study on consensus approaches for conventional GDM and LGDM was undertaken by Labella et al. [75]. The key question of their study, which gives its title to the article, is whether classical consensus approaches can suitably handle CRPs in LGDM. A major goal of their investigation consists in further reflecting on some previously identified challenges faced by classical approaches to cope with large groups [108], such as: temporal cost, constant supervision of preferences, complexity of dealing with non-cooperative behaviors, etc. The main conclusion drawn is that whilst some classical consensus models can be easily enhanced to deal with large groups, other models for specific decision frameworks are more complex to escalate, and so LGDM-specific versions should be developed to translate these frameworks into new ones to accommodate larger groups. Some typical requirements to do this are listed, e.g. weighting the relevance of alternatives in the consensus measures for models based on feedback mechanisms, enhancing models based on automatic adjustments to take into account multiple experts in the same round simultaneously, and making optimization approaches more flexible. Additionally, depending on the cohesion or diversity of the group, the discussion in [75] suggests that contemporary consensus models for LGDM ought to flexibly adapt to the use of feedback or automatic adjustments, or even seek the appropriate balance between both, as in [104].

4.5 Behavior Modeling and Management

Surveyed works on behavior modeling and management are summarized in Table 4.5 and subdivided in the following four themes:

- Detecting and penalizing uncooperative behaviors in CRPs.
- Managing minority opinions and uncooperative behaviors.
- Self-management and mutual evaluation mechanisms for behavior management.
- Analyzing diverse behavioral styles.

4.5.1 Detecting and Penalizing Uncooperative Behaviors in CRPs

Palomares et al. exhaustively investigated the process of detecting, quantifying and penalizing non-cooperative behaviors in CRPs for LGDM, with several published results [105, 106, 109]:

Table 4.5 Summary of surveyed works on behavior modeling and management

Theme	Representative publications (listed by theme alphabetically)
Detecting and penalizing uncooperative behaviors in CRPs	Palomares et al. [105, 106, 109]; Shi et al. [130]; Zhang et al. [187, 192]
Managing minority opinions and uncooperative behaviors	Xu et al. [165]
Self-management and mutual evaluation mechanisms for behavior management	Dong et al. [36, 38]
Analyzing diverse behavioral styles	Carneiro et al. [22]

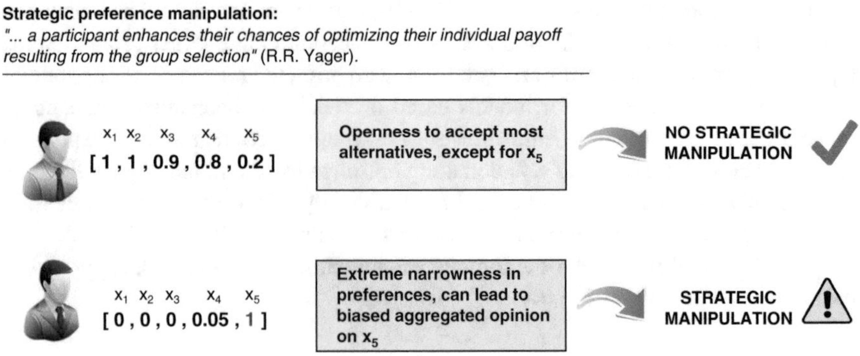

Strategic preference manipulation:
"... a participant enhances their chances of optimizing their individual payoff resulting from the group selection" (R.R. Yager).

| | x_1 x_2 x_3 x_4 x_5 [1 , 1 , 0.9 , 0.8 , 0.2] | Openness to accept most alternatives, except for x_5 | NO STRATEGIC MANIPULATION ✓ |

| | x_1 x_2 x_3 x_4 x_5 [0 , 0 , 0 , 0.05 , 1] | Extreme narrowness in preferences, can lead to biased aggregated opinion on x_5 | STRATEGIC MANIPULATION ⚠ |

Fig. 4.14 Example illustrating the existence of strategic preference manipulation investigated by Yager in [173], on numerical unit-interval preference vectors over $n = 5$ alternatives

• Palomares et al. introduced in [105] a methodology that (to the author's knowledge) constitutes the first research effort focused on managing non-cooperative behaviors in CRPs conducted by large groups, thereby founding the main ideas and principles for all the subsequent works in this trend. The methodology is motivated by Yager's state-of-the-art work to identify and penalize strategic preference manipulation in classical GDM processes without CRPs [173], whose motivation is graphically illustrated in Fig. 4.14. A consensus model equipped with an increase-decrease feedback mechanism and based on pairwise similarities between experts for consensus measurement, is extended as shown in Fig. 4.15. A distinctive novelty of this consensus model is the integration of a Fuzzy C-Means (FCM) based clustering process to divide the large group into fuzzy clusters, so that each member might belong to multiple clusters with different levels of membership in the [0, 1] interval. After clustering, a set of rules based on cluster and preference analysis between two consecutive CRP rounds are defined to detect not only non-cooperative individuals (also referred to as *outliers*), but also non-cooperative subgroups or coalitions. Furthermore, non-cooperative behaviors detected can be applied either a full weight penalization or a partial weight penalization scheme. The difference between full and partial penalizing resides in whether the aggregation of similarities for measuring consensus considers the (penalized) weights of uncooperative experts or not. Through a case study conducted in the province of Jaén[3] (Spain), it is demonstrated that the extended consensus model for LGDM can effectively detect non-cooperative behavior, and their penalization contributes to a better quality and more efficiently achieved consensual decision. Despite its novelty and remarkable effectiveness in detecting individual and coalition non-cooperative

---

[3] **Author off-topic note**: Jaén is a province situated in the south of Spain, one of the eight provinces forming the region of Andalucía, and the book author's natal province. Discover more in the English website for Jaén province: http://www.andalucia.org/en/destinations/provinces/jaen/.

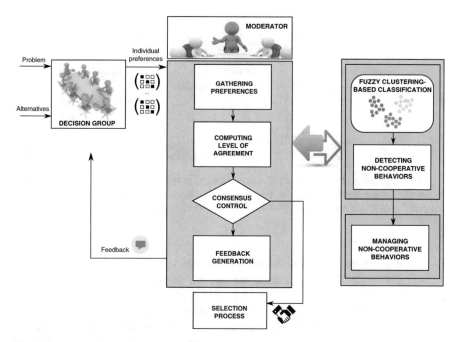

Fig. 4.15 Extended consensus model with a fuzzy clustering-based approach to detect and manage non-cooperative behaviors in LGDM (adapted from [105])

behaviors, three limitations of the model are (1) the necessity of determining the number of clusters K in advance, owing to the nature of the FCM algorithms, (2) the effort invested by the human moderator in managing behaviors in the cases where a full implementation of the model is not available, and (3) the impossibility to "recover" the importance weights of experts who are initially uncooperative but they rectify their behavior later on throughout the CRP.

- In [106] Palomares et al. presented their second contribution on behavior management for LGDM, in a linguistic preference relation setting. Unlike [105], the weight penalization applied at a given CRP round t is not propagated to the subsequent round $t + 1$. Therefore, the importance weight of a non cooperating expert can increase again if his/her level of cooperation improves after receiving additional feedback to adjust his/her opinions. In summary, the importance weights are re-calculated at each CRP round, independently of the values of such weights at previous CRP rounds. A cooperation coefficient is introduced and measured for each expert, based on principles from fuzzy set theory and computing with words [176, 181]. The coefficient is formally defined as a function of the proportion of suggested increase-decrease feedback on assessments that the expert accepts to apply at the tth round, out of the total number of advices received. To ensure a proper convergence speed towards consensus, the semantics of cooperativeness given by a fuzzy set become stricter as the CRP advances (see Fig. 4.16). An application example is undertaken by a company for

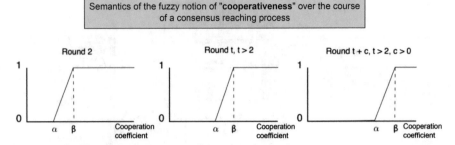

Fig. 4.16 Evolution of the semantics of cooperativeness over the course of a CRP: as the number of rounds undertaken increases, the notion of cooperativeness becomes stricter, i.e. a larger cooperation coefficient is required to deem the behavior of experts as cooperative (adapted from [106])

selecting the most accepted charity action of the year. In spite of overcoming the previous limitation by facilitating the recovery of previously penalized weights, the cooperation coefficient defined in this model only considers the experts' behavior in the present CRP round t, ignoring the previous history and evolution of each participant's behavior. This would be an important aspect to consider in situations when two experts have a similar cooperation coefficient at the current round, but they exhibited different behaviors at previous rounds.

- In [109] the authors take an step forward and extend the "single-round" cooperation coefficient from [106]. Their study enables the analysis of the overall behavior *trend* adopted by experts across the CRP, thereby taking the *evolution* of such behavior into account. The principles from dynamic MCGDM, previously investigated by Campanella and Ribeiro in [20] and Zulueta et al. in [195], are adopted in their work as follows: given a CRP round t and an expert $e_i \in E$, two indicators describing her/his present (round t) and past behavior (previous rounds $t' < t$) are aggregated by using Yager and Rybalov's uninorm operators [171]. The associativity and full reinforcing properties of uninorm operators allow, respectively, to: (1) recursively incorporate information about the past behavior of the expert, and (2) emphasize the overall cooperation coefficient (and thus the resulting importance weight) of experts who continuously and consistently exhibit a cooperative or non-cooperative behavior across several consecutive CRP rounds. As a result, a more realistic measure of cooperativeness is devised, while convergence towards consensus remains higher when patterns of non-cooperativeness are detected and dealt with.

Zhang et al. [187] tackled a more complex and highly diverse LGDM framework, where participants provide heterogeneous preferences under either one of the following six types: ordinal, numerical, interval-valued, fuzzy numbers, linguistic terms or Intuitionistic Fuzzy Numbers. The heterogeneous information is firstly unified into preference relations and then transformed into priority vectors of assessments on alternatives, using the OWA operator. An aggregation of Individual

Priority (AIP) mechanism is used to prevent loss of information during the unification process. Based on the unification approach, a consensus model for LGDM with heterogeneous information and capable of managing different levels of information granularity, is proposed. The consensus model is characterized by its revision schemes to penalize uncooperative behaviors during consensus reaching. The temporary collective preference utilized for guiding consensus is based on non conflicting opinions exclusively and, unlike most feedback generation mechanisms, in [187] the advice generated is generic and provided to the entire decision group, instead of producing individual personalized feedback. This is motivated by the difficulty of providing individual feedback to every member of a very large group. The classification of behaviors is as follows: those participants who apply the generic feedback (and originally required to), are deemed as cooperative, and vice versa. Non cooperating individuals are then penalized accordingly.

The recent work by Shi et al. in [130] constitutes another novel step in continuing the above described research initiated by Palomares et al. [105, 106, 109]. Their primary extension consists in comprehensively analyzing—for each expert— not only the patterns of uncooperative behavior, but also possible patterns of positively cooperative behavior in parallel. Specifically, the authors argue that when an expert is required to adjust multiple assessments of her/his decision matrix at given consensus round, it may occur that: (1) the expert exhibits an inherently cooperative behavior or *cooperative-leading behavior*, bringing all the identified assessments closer to the consensus opinion; (2) the expert exhibits an inherently uncooperative behavior or *non-cooperative leading behavior*, keeping all the identified assessments unchanged or even moving them against consensus; (3) the expert shows a mixed or *average behavior*, acting cooperatively in some of her/his identified assessments whilst remaining uncooperative in other assessments identified at the same CRP round. This motivates overcoming an existing limitation in previous approaches: the behavior of experts is deemed as single-dimensional, i.e. it is either cooperative or uncooperative, whereas in practical situations it may occur that the same expert exhibits different degrees of cooperativeness and non-cooperativeness simultaneously. Shi et al. propose a solution characterized by using uninorm aggregation operators, as follows (see Fig. 4.17):

1. A degree of non-cooperativeness, $\#NCOOP \in [0, 1]$ is calculated for each expert and consensus round, similarly as in previous behavior management models [106, 109].
2. In parallel to the calculation of $\#COOP$, it is also computed a degree of (positive) cooperation, $\#COOP$, predicated on the preference adjustments made in accordance with the advice received.
3. Once calculated $\#COOP$ and $\#NCOOP$, they are combined to obtain a comprehensive cooperation coefficient $CCOOP = \mathcal{U}(\#COOP, 1 - \#NCOOP)$, with $\mathcal{U}$ a uninorm aggregation operator (see Chap. 2, Sect. 2.3). Notice that the aggregation process actually fuses the degree of positive cooperation with the *absence* of non-cooperative behavior, given by $1 - \#NCOOP$, in order to describe the comprehensive cooperation level of the expert. The full reinforce-

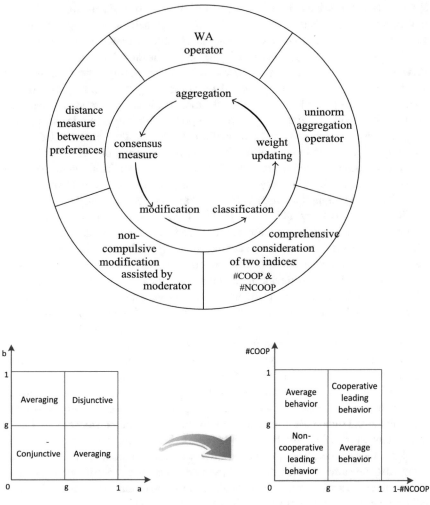

Fig. 4.17 Uninorm-based consensus model for comprehensive behavior classification presented by Shi et al. [130]

ment property of uninorm operators is manifested in the following two particular cases:

- *Upward reinforcement*: The expert shows a high level of positive cooperation and a low level of non-cooperation, and thus both aggregation inputs are high.
- *Downward reinforcement*: The expert shows a low level of positive cooperation and a strong level of non-cooperation, and thus both aggregation inputs are low.

As Fig. 4.17 shows, each of three possible adopted attitudes adopted by the uninorm operators implies one of the three types of behavior previously obtained, i.e.

cooperative-leading vs *non-cooperative leading* vs *average* behavior. To demonstrate the added value and efficiency of the proposed consensus model, three baseline consensus models are presented for comparison through an illustrative example of MCLGDM problem that includes the 2-D visualization of experts' preferences throughout the CRP [107].

Zhang et al. [192] investigated the mutual influence between Social Network Analysis (SNA) and the existence of non-cooperative behaviors in CRPs. They developed a SNA-based consensus approach that takes non-cooperative behaviors of participants into account in a social network MCLGDM context. Unlike previous similar studies that deem the trust information across the social network structure as static, here the experts might modify their trust values towards other known participants, influenced by their (sometimes non-cooperative) behaviors. Therefore, the experts not only provide decision matrices regarding a set of alternatives and criteria, but they also supply trust values for other experts in their social trust network. Although naturally experts would not be able to assess their trust towards every member in the group and, as a result, some trust values between pairs of experts would be initially unavailable, a complete pairwise trust model is yielded through a trust propagation and aggregation process on the explicitly provided trust values: (1) incomplete trust information is firstly yielded via trust propagation, by using a t-norm operator $T(\cdot, \cdot)$ as illustrated in Fig. 4.18, (2) all trust degrees towards a given expert are subsequently combined using an averaging aggregation operator such as OWA. The importance weights of experts are determined upon the complete social trust information and embedded into the CRP. During the CRP, the experts not only adjust their decision matrices to achieve the required consensus level, but they can also update the trust information initially provided towards other known experts in the social network. This is done in accordance of a non-

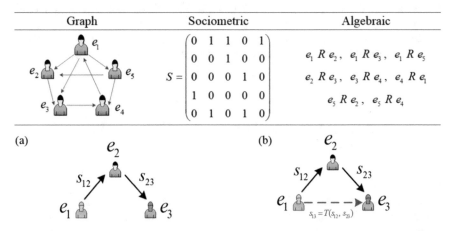

(a)

(b)

Fig. 4.18 Top: different representation formats for a social trust network. Bottom: example of trust propagation process to determine the trust degree from e_1 to e_3 [192]. (**a**) No direct trust between e_1 and e_3. (**b**) Trust propagation between e_1 and e_3 via e_2

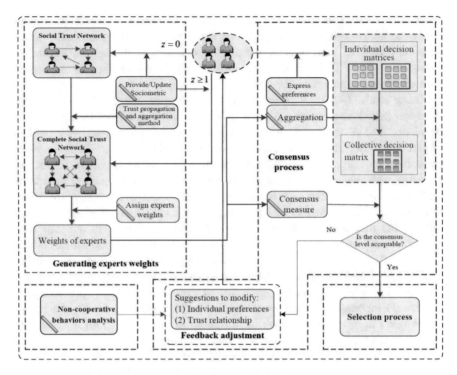

Fig. 4.19 SNA-based consensus framework with behavior management and expert weighting upon dynamic social trust information [192]

cooperative behavior identification process undertaken during the CRP, such that the weights of the experts exhibiting non-cooperative behaviors would be decreased. The overall approach, depicted in Fig. 4.19, constitutes the first consensus support model for social network-driven MCLGDM that meaningfully integrates dynamic inter-user trust information, multiple patterns of non-cooperative behaviors and expert weighting.

4.5.2   Managing Minority Opinions and Uncooperative Behaviors

Critical consensual decision making situations should not only monitor non-cooperative behaviors, but also detect and deal with minority opinions that may sometimes constitute valuable decision inputs not to be ignored (and other behavior management models fail to consider). This is one of the objectives in the work by Xu et al. [165], which revolves around a number of principles:

1. Most consensus models for LGDM fail to consider the time pressure and complexity inherent to emergency MCLGDM situations (e.g. a flooding incident in a coal mine).
2. A central aspect in such MCLGDM situations is to strike a balance between quality decisions (as a result of an effective behavior management) and timeliness (to enable efficient and optimal emergency response).
3. The influence of majority opinion clusters might sometimes lessen the presence of minority positions, hence attention should be paid to identifying and properly treating these minorities.

Thus, the novelty in [165] lies in handling non-cooperative behaviors in a time-sensitive manner, whilst preserving minority opinions and avoiding their—sometimes accidental—loss of information in the overall decision making procedure. Different types of minority opinions are described in the study, outlining their main characteristics. Preferences are given by numerical decision matrices under different scales and distinguishing among cost and benefit criteria, therefore the decision matrices are firstly normalized before clustering the large decision subgroup into subgroups (we refer the reader to Section X for clustering-focused methods by Xu et al. and other authors). Based on subjective and objective information, a comprehensive adjustment coefficient is also introduced and utilized in the feedback generation process for building consensus.

4.5.3   Self-management and Mutual Evaluation Mechanisms for Behavior Management

Dynamic and self-managing approaches for handling non-cooperative behaviors were recently presented by Dong et al. [36, 38]:

- Dong et al. [36] presented an insightful mechanism in which experts' knowledge about each other plays a key role in the management of non-cooperative behaviors. Besides a preference relation, each participant supplies a Multi-attribute Mutual Evaluation Matrix (MMEM) containing assessment information about other known experts' features, e.g. professional skills, cooperation and fairness. Through an optimization model, the current mutual evaluation information at a given CRP iteration is used to calculate the importance weight assigned to each expert. However, the most remarkable contribution in Dong et al. study is arguably its behavior management mechanism. Three different types of non-cooperative behavior are defined to accurately reflect the diversity of participants' profiles inherent to a LGDM problem. The degree to which these types of behavior are exhibited by experts is calculated. Based on some hypotheses on the effect of these behaviors on the MMEM attributes, the mutual evaluation assessments are then adjusted by the group members. A visual summary of the types of behaviors considered in this work and [38], alongside their relationship

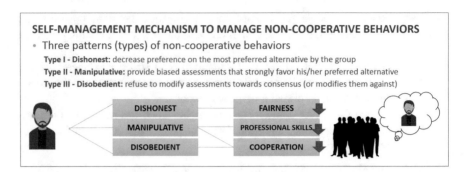

Fig. 4.20 The three types of non-cooperative behaviors investigated by Dong et al. in [36, 38] and their influence in the attributes for mutual expert evaluations

with MMEM attributes, is provided in Fig. 4.20. Consequently, an update process is undertaken on their importance weights and those from their peers. Simulations for a large number of experiments show the reduction on the average number of rounds required to achieve consensus, as well as an increase in the ratio of successful CRPs, i.e. achieving consensus within a predefined maximum number of rounds.

- Subsequently, Dong et al. argued in [38] that having a human moderator responsible for overseeing CRPs for LGDM is an excessively demanding endeavor. They present a framework for CRP in LGDM, based on a mechanism that self-manages different types of non-cooperative behaviors. Experts are classified into subgroups using a clustering technique and they provide MMEMs for members in their subgroup. Similarly to [36], three types of non-cooperative behaviors are considered and referred to as type I, type II and type III behaviors. Unlike previous works from other authors where non-cooperative behaviors are dealt with similarly and under a single overall mechanism, in [38] each type of behavior is identified and dealt with differently, associating them to distinct human participant attributes amongst the ones used in the MMEMs. However, the main difference with [36] is that experts are only requested to evaluate the attributes of participants in the same subgroup, thereby making this task much more manageable for them. A simulation analysis and comparison against other consensus models for LGDM is presented and discussed. The added value of this model is the smaller human effort required in providing peer evaluation information—done as smaller subgroup level—and the alleviation of the moderators excessive monitoring tasks. On the contrary, simulations present the limitation of not covering all possible realistic settings exhaustively.

Interestingly, the feedback mechanism in the two above studies suggests experts to bring their assessments within the closed interval between their current assessment and the collective assessment (or vice versa), explicitly allowing them the freedom to modify the identified assessment to a variable degree, or even not modifying it.

4.5.4 Analyzing Diverse Behavioral Styles

In [22], Carneiro et al. investigated the use of agent modeling for mirroring—and acting according to—the intentions of human decision makers in a LGDM process. The agent-based modeling relies on five distinct behavioral styles (*Dominating, Integrating, Compromising, Obliging* and *Avoiding*). This allows agents to act in accordance with the intentions of the decision makers on behalf of whom they act. The behavioral styles are in turn defined in accordance with four relevant dimensions in GDM and LGDM: (1) concern for oneself, (2) concern for others, (3) resistance to change and (4) activity level. To study the research hypothesis formulated in their work, the authors devised a multi-agent system that incorporates an argumentation-based negotiation model for seeking collective consensus. A comprehensive study that simulates similar interactions to face-to-face meetings with up to 40 participants completes the paper, showing that group decision support systems should allow participants to express different intentions through behavioral styles, since important LGDM processes may span a certain period of time. The best decision results are achieved when individuals select the behavior style that best represents their true intentions, and an agent-based decision support tool may use this information algorithmically to seek a consensual, high-quality decision.

4.6 Theory and Interdisciplinary Approaches

Table 4.6 summarizes the main theoretical aspects and related disciplines to each of the studies reviewed in this section.

Rodríguez presented two theoretical-practical works on collective intelligence systems and their dynamics at scale [116, 117]:

- Firstly, in [116] Rodríguez introduced a novel conceptual framework for creating collective intelligence systems at societal scale, as well as a methodology for implementing societal decision support systems for LGDM predicated on social network theory. Two arguments are postulated on how the envisaged kind of system would differ from a traditional GDSS: (1) a LGDSS should be able to deal with problems involving highly fluctuating group participation, (2) conventional GDSS focused on small groups may fail to capture the heterogeneity of participating individuals. As a result, the framework proposed in [116] considers a general-purpose system able to tap collective intelligence at a societal scale, underpinned by the use of social network topologies to provide representations of group opinions.
- Secondly, the author proposed in [117] another methodological approach for implementing social DSS, characterized by aggregating perspectives of a diverse group in a computer-mediated setting. The social decision making process is described as consisting of three major stages: individual solution ranking, collec-

Table 4.6 Summary of surveyed works on theory and interdisciplinary approaches

Publication	Theoretical aspects	Related discipline(s)
Back et al. [5]	Psychological attractiveness towards preferred alternative	Neuroscience
	Discordance between preferred alternative and decision made	Cognitive sciences
Goel and Lee [48]	Scaling deliberative and participatory democratic processes	Democracy and political sciences
	Triadic decision making	Behavioral sciences
Nyerges and Aguirre [98]	Large-scale GIS	Geographical sciences
	Geo-visual analytics	
Rodríguez [116, 117]	Collective intelligence systems	Social sciences
	Social network analysis	Democracy and political sciences
Shum et al. [131]	Interaction and argumentation	Social sciences
	Social network analysis	Argumentation theory
Xue and Xu [169]	Coordination	Construction management
	LGDM under diversity and complexity	Project management

tive solution ranking, and selection. Proxy-based representations are adopted to accommodate fluctuating participants. The author also proposes incorporating a suite of vote aggregation approaches (e.g. weighted voting algorithms) to enable their application in diverse real-life political systems (e.g. direct or representative democracy, dictatorship, etc.). This is one of the earliest approaches in LGDM literature that incorporates social trust information stemming from a social network. The social network is in turn modeled by a weighted multi-relational network connecting experts, problem domains and solution domains in a semantic fashion, similarly as an ontology.

Back et al. presented in [5] a LGDM study from a neuroscience and cognitive sciences perspective. The authors contend that after a decision is made and where relationships exist between present and future decision making processes, some participants tend to modify the psychological attractiveness of the alternatives in their favor. In particular, it is shown that individuals in large groups tend to change the attractiveness of their preferred alternative between the pre-decision and the post-decision stages. The claim is made as a result of a experimental study in a setting where opinions are expressed as fixed-scale ratings on alternatives, and a ballot voting process is conducted followed by a majority, in-group authority or out-group authority rule to select an alternative. Furthermore, if there exists some discordance between the decision made and the own preferred alternative of a participant, (s)he will often lean towards further reinforcing her/his attractiveness towards the individually preferred alternative. Back et al. also argue that the perception of a decision procedure and its outcomes as *fair* promotes a stronger sense of acceptance towards the decision made. As stated by the authors of the study, "in large groups, many decisions made (will) affect all members of the group" [5].

In [98], Nyerges and Aguirre presented an evaluation of the quality and scale of interactions in a large-group field experiment conducted on 179 participants by using an online public participatory geographic information system (GIS), whereby questionnaire information is used to collect the opinions of participants. The main research question concerns whether the aforesaid decision process would benefit from a platform endowed with GIS technology. The research concludes that geo-visual analytics tools such as the proposed tool called *Grapevine*, helps identifying productive clusters such that deliberative processes and discussions in them become easier. These decision support tools are also proved to help shifting opinions towards agreement, namely on an application scenario on regional package transportation. An architectural description of the proposed GIS-based system is provided in Chap. 5.

A perspective on using social network analysis for supporting collective decision making at scale was provided by Shum et al. in [131]. They considered scenarios with multidisciplinary working groups (e.g. multiple firms or multiple departments within a firm) whom each have diverse perspectives, different individual knowledge and distinct approaches to solve a problem. The Internet-related technologies allow applications to make effective and efficient decision making through deliberation, accessing a wealth of information and fostering a more rapid, deeper and less costly dissemination of knowledge, even if subgroups or stakeholders are geographically separated. Similarly to state-of-the-art opinion dynamics models, the individual opinions in this study (text-based and modeled as arguments) are assumed to change as a result of interaction and argumentation among participants. Argumentation is described as a tool for collaborative distributed decision making, helping groups to give an insight of the way real-world social decision making processes would take place under a non-competitive setting. The work also explores how collective argument analysis can be used to investigate structural properties of interacting social networks.

Goel and Lee [48] conducted a study focused on how to scale participatory and deliberative democratic processes. Whilst deliberative processes and participatory practices are reported to foster highly engaged citizens, scaling such practices up to thousands or millions of people is unquestionably a major challenge in terms of increasing and maintaining their participation, largely due to cognitive and psychological barriers which in very large groups hinder the willingness to participate. In other words, humans tend to become more disengaged when being part of a large group. A major finding in [48] consists in demonstrating the importance of facilitating small group decisions based on triads (groups of three), by decomposing the LGDM problem into sequences of small group interactions, which take place separately but intertwined, forming a graph. Put another way, "the deliberative decision-making process can be, at its core, the interaction of many separate but intertwined deliberations" [48]. The proposed triadic decision making procedure with majority rule is depicted in Fig. 4.21 and summarized below:

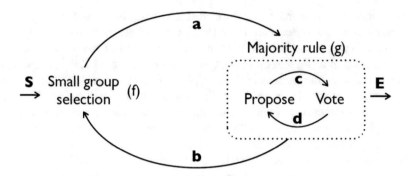

Fig. 4.21 Triadic decision making procedure proposed by Goel and Lee (source: [192])

(S) First, a small group is selected (f).
(a) Once a small group of three has been selected, it initiates a majority rule process that commences with a propose action.
(c) All three participants in the triad vote on the proposal made.
(d) It is required that at least two out of the three participants vote for the proposal, otherwise another proposal is taken and the triad moves back to (c).
(E) If two of the three participants accept the proposal and one participant motions to end the triadic process, then the decision process ends.
(b) Otherwise another small group is selected going back to (f).

Overall, the study shows that only a small number of triadic units like the one outlined above are required to achieve feasible and scalable deliberative decision making on a participatory fashion. However, it is impossible to protect against extreme outliers, e.g. arbitrarily unreasonable individuals whose opinions and conduct may affect the decision made.

More recently, Xue and Xu [169] discussed the LGDM-challenges in construction management projects for major infrastructures, e.g. for the sustainability evaluation of such major infrastructure projects. Their contributions include: (1) a discussion of ideas for an innovative conceptual framework based on dynamic interactive coordination, aimed at determining suitable groups of experts, eligible attributes and their importance weights, as well as integrating preferential information in major infrastructure project sustainability evaluation problems; (2) a reflective analysis on the processes of measuring interactive coordination and effective integration of collective decision information, with particular focus on their impact on the life-cycle nature of construction management processes; (3) a set of guidelines on key assessment indicators to cope with the target domain. The study emphasizes the high complexity of the application scenario, involving considerable diversity among experts across multiple layers or roles. The importance of building consensus and the consideration of feedback and opinion supervision mechanisms are raised as lines for future work in this domain.

Acknowledgments The author and contributors of this chapter would like to thank those colleagues across the LGDM and related scientific communities who willingly shared their original graphical material during the elaboration of the literature survey, specially to: **Yucheng Dong** (Sichuan University, China), **Bingsheng Liu** (Chongqing University, China), **Victoria López** (Complutense University of Madrid, Spain), **Ramón Soto** (Sonora University, Mexico), **Xunjie Gou** and **Francisco Herrera** (University of Granada, Spain), **Ashish Goel** and **David T. Lee** (Stanford University, US).

Chapter 5
Implementations and Real-World Applications of LGDM Research

Abstract Due to their rather practical nature, existing works on real-world implementations of Large Group Decision are summarized in this chapter, along with a brief overview of the real-world practical scenarios where many of the surveyed studies have been applied.

5.1 Large Group Decision Support Systems

This section focuses on describing some examples of software implementations for LGDM and consensus models into DSS.

5.1.1 Social LGDSS

One of the first DSS implementations specifically envisaged to support large decision groups was arguably presented by Turoff et al. in [146], as part of their efforts to address the fragmentation and diversification of society with a continuously increasing number of interest groups. As such diversification impedes the ability to foster consensual collective decisions at a large scale, Turoff et al. developed a social DSS to support various decision frameworks with thousands of participants on complex topics and under the presence of diverse (and frequently opposing) views. The main objectives of their social DSS are enumerated as follows:

1. To facilitate the integration of diverse views into a growing knowledge base.
2. To expose and consider all possible alternatives to make a decision on a given societal topic, and their underlying rationale.
3. To facilitate the emergence of new options halfway in the decision process.
4. To reflect on the previous history of decision processes and results, and identify suitable people to contribute to specific issues accordingly.

© The Author(s), under exclusive licence to Springer Nature Switzerland AG 2018

I. Palomares Carrascosa, *Large Group Decision Making*, SpringerBriefs in Computer Science, https://doi.org/10.1007/978-3-030-01027-0_5

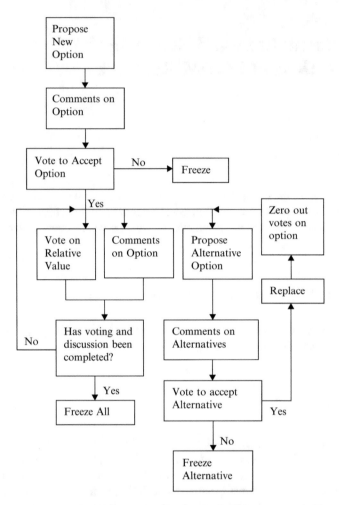

Fig. 5.1 General workflow for the implementation of a Social DSS (source: Turoff et al. [146])

The central application example for the proposed LGDSS concerns democratic representative processes, in which individual opinions are defined by choice of single options from a non-predefined set, and where voting methods are used to aggregate these opinions (Fig. 5.1).

5.1.2   LaSca

A few years after Turrof et al.'s proposal, Carvalho et al. introduced in [24] LaSca: a system for supporting large-group decisions. Inspired by diverse decision making theories, LaSca allows its users to flexibly define the structure of the large group formed by them: as the authors suggest, the system helps users "deciding how to decide" [24]. Before presenting the system, Carvalho et al. provide in their paper an interesting motivation for LGDM: deciding means choosing one or more alternatives out of those available to achieve a goal sought, and in order to minimize problems stemming from bad decisions, LGDM becomes indispensable to incorporate individuals with different knowledge or perspectives about the problem domain at hand. As mentioned above, a key problem to be addressed is how to structure and better manage the participation of the entire group throughout the decision making process. Complexities such as reuniting all participants in the same venue, finding a suitable schedule to everyone, the duration of LGDM sessions, and the loss of encouragement by some participants, are discussed as "obstacles" against undertaking a proper large-group session. Intuitively, Internet access overcomes several of these limitations and also enables asynchronous participation. LaSca approach has much in common with the idea of participatory design exposed by Muller [90], which includes an initial exploration and exchange of ideas to come up with a shared understanding of the problem.

In essence, LaSca users can choose which well-known decision theory (or combination of them) is the most suitable to tackle a specific problem. The system complies with a classical client-server architecture under the Model-View-Controller architectural pattern and subject to a single-point failure approach, such that any technical failure in a participant's side will not affect other participants. Figure 5.2 shows two examples of LaSca user interface. Three user roles are incorporated: participants, creator and moderator. The creator of the problem, for instance, would be usually in charge of defining the appropriate problem structuring.

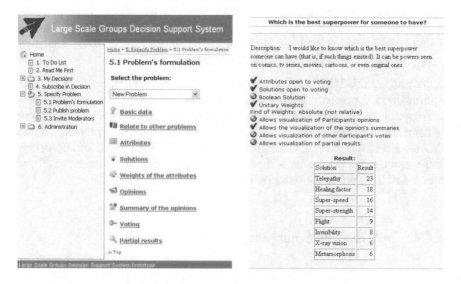

Fig. 5.2 Examples showing LaSca user interface (source: [24])

The moderator is able to elicit and analyze responses (preferences) supplied by participants, and likewise identify situations such as the presence of repeated or irrelevant information.

5.1.3 MENTOR

A visual monitoring tool called MENTOR for analyzing the evolution of LGDM processes was introduced by Palomares et al. in [107]. Based on Kohonen's Self-Organizing Maps (SOMs) [74] for the low-dimensional representation of high-dimensional individual preferences, MENTOR is envisaged to assist LGDM process analysts in monitoring the state of ongoing LGDM problems under different contexts, particularly those involving a CRP. Whilst the tool is presented under the context of preference relations and these are encoded as numerical preference vectors for applying the SOM algorithm, it can be seamlessly extended into other preference structures and assessment formats with the aid of unification approaches [28]. The training data describing individual and collective preferences are used to self-train a SOM structure for visualizing the problem at its current stage, and identifying a variety of important aspects, including:

1. The relative position of individual preferences with respect to each other and with respect to the collective (subgroup or large group) aggregated preferences.
2. The identification of trends, i.e. subgroups of a considerable size with very similar opinions to each other.
3. The identification of individuals with drastically different opinions to the rest of the group.
4. The evolution of the opinions of participants over the successive rounds of a CRP.
5. The visual identification of non-cooperative behaviors at individual and coalition level, enabling a visual aid that can be combined with existing analytical approaches for behavior detection, e.g. [105].

Importantly, MENTOR aims at providing a "visual" alternative to traditional information systems provided by other DSS for analytical purposes, most of which are based on purely textual or numerical information. While the human-led analysis of numerical/textual information is feasible in small group decision problems, it may become a daunting task in a LGDM context where a high volume of information about preferences, consensus and proximity degrees, etc., must be analyzed and interpreted in a time-sensitive manner (Figs. 5.3 and 5.4).

5.1.4 Web Tool for Emergency LGDM

Xiang [91] developed a Web tool for LGDM on energy network dispatch optimization under emergencies of local energy storage. In emergency LGDM, large group decisions should be made within a very limited amount of time. Similarly

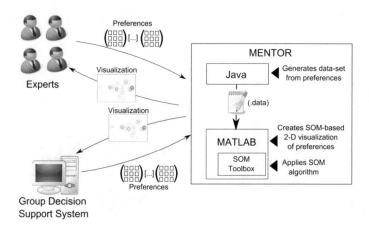

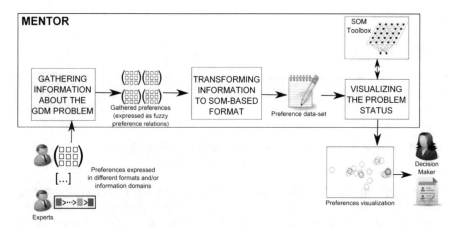

Fig. 5.3 Architecture and operation scheme of MENTOR (source: Palomares et al. [107])

to other works, Liu expounds that existing LGDM methods are time consuming, excessively impracticable and costly in the aforesaid scenarios. The proposed solution revolves around a Web platform, whose ubiquitous capabilities naturally allow physically separated experts to log in and provide their preferences (given by discrete numerical decision matrices) anywhere. Various functionalities are integrated to meet the cornerstone goals of the system:

1. Finding a discrete finite set of alternatives within the domain of emergency local energy shortage, under the existence of uncertainty.
2. Facilitating the elicitation of preferences from a large group.
3. Managing large-group CRPs to select the most acceptable alternative in the previously identified set, in an automatic and effective manner. Adjustments on assessments are made via an integrated optimization model.

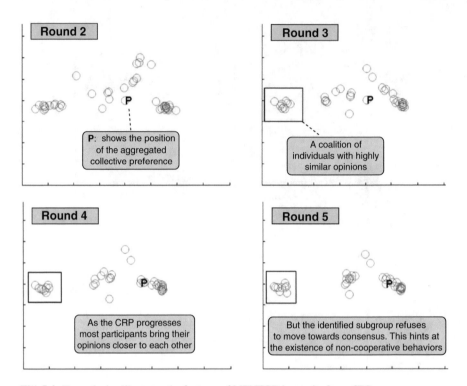

Fig. 5.4   Example that illustrates the features of MENTOR in monitoring a CRP

The large group is a priori clustered into subgroups, by associating each subgroup with a local network in the Web tool, e.g. multiple participating stakeholder companies. The associated publication [91] provides an overview of the database design for the Web DSS. Its architecture is illustrated in Fig. 5.5.

5.1.5   COMAS (COnsensus Multi-Agent System)

The semi-supervised agent-based consensus model introduced by Palomares and Martinez in [104] and reviewed in the previous chapter, was developed along with a multi-agent based platform to enable distributed CRPs at scale via a Web interface [102]. The system consists of two interfaces: one for the moderator or organizer of the LGDM problem, and one for experts to provide their relevant information. The expert interface is depicted in Fig. 5.6, where each expert inserts their opinions as additive preference relations on the set of alternatives registered by the problem organizer a priori. Furthermore, each expert chooses her preferred change profile (see Fig. 4.9 in Chap. 4), i.e. the attitude towards preference changes used to instantiate an autonomous agent that emulates the human expert actions to apply some of the required modifications on preferences.

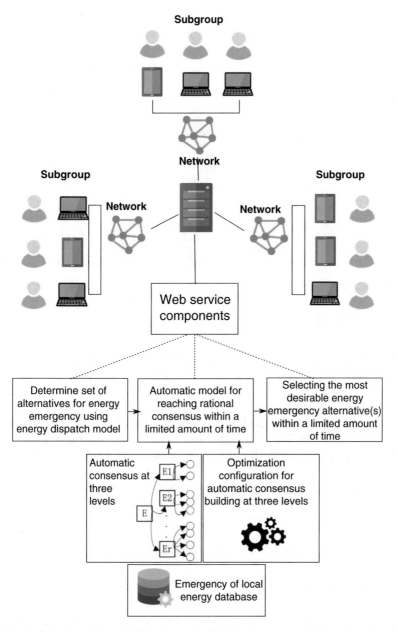

Fig. 5.5 Architecture of the web tool proposed in [91] for energy network dispatch optimization under emergencies of local energy storage

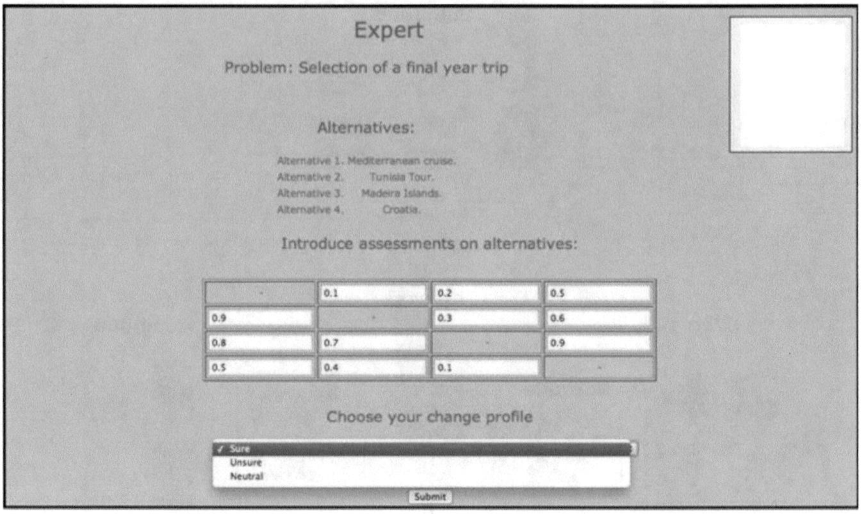

Fig. 5.6 Expert interface of the semi-supervised multi-agent based web platform [102]

5.1.6 Multi-Agent System for Scalable GDM

Prior to the consensus-driven platform developed by Palomares et al., Husain in [64] proposed a system to support group decisions in collaborative environments, also based on a multi-agent approach. Husain's multi-agent system is inspired by the behavior of swarms in nature, e.g. honeybees. By simulating patterns of swarm behavior, the proposed solution is reported as reliable and fault tolerant, as opposed to centralized approaches. A set of mobile agents are created for performing multicast group communication and GDM processes in a scalable manner. Each participant modeled by a mobile agent proposes a solution, and migrates to collect other participants' proposals, fostering the exchange of proposals and selection of a final one. This highly interactive process is represented in Fig. 5.7.

5.2 Practical Applications of LGDM

The chapter concludes with a listing of popular real-world applications where the surveyed LGDM studies and developments have focused to date. The list shows the relevance and necessity of making large group decisions in an ample variety of domains, thus motivating the need for further applied research developments in this area along with related ones. Some of the cited works below are not included in the previous survey due to having focused on LGDM or MCGLDM at the time of writing, but their extensions into such decision contexts is particularly encouraged.

- *Project management*: Project evaluation and selection [193], contractor selection [82].

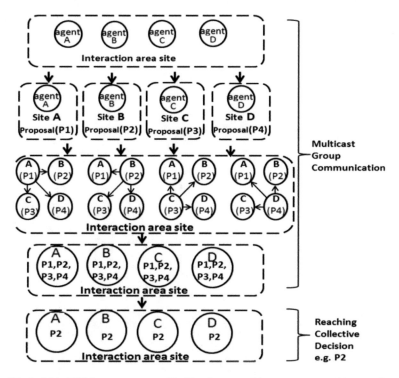

Fig. 5.7 Scalable GDM process proposed by Husain as a multi-agent system architecture (source: [64])

- *Energy and Environmental Sciences*: Hydropower plant construction planning [81], water management [135], green energy supplier evaluation [151], environmental impact evaluation [182], degradation control in rivers [185].
- *Marketing*: New product development [29, 37, 174], new product evaluation [138], customer service selection [78].
- *Finance and banking*: Loan evaluation and risk analysis [13].
- *Emergency decision making*: Fire mitigation [19], energy network dispatch [91], rescue plan choice [155, 162], flooding accident management [165], earthquake shelter selection [168].
- *Human Resources*: Employee award selection [36], academic salary reform [86], academic recruitment [87, 152, 175].
- *Political Sciences*: Participatory and deliberative democracy [48], representative democracy [146], decentralized voting [32], e-government [116].
- *Systems Reliability*: Failure Mode and Effect Analysis [80].
- *Charity*: Selection of charitable actions [109].
- *Leisure and tourism*: Group trip selection [104], restaurant selection for corporate dinner [154], recommendation of activities for large groups [43].
- *Artificial Intelligence and Robotics*: supporting consensus decisions in swarms of robots or autonomous agents [33].

Chapter 6
Conclusions and Future Directions of Research

Abstract This chapter concludes the book by summarizing the main conclusions derived from the LGDM research advances and pending challenges to date. Some proposed directions for future research in this topic are finally highlighted.

6.1 Conclusions

GDM problems and consensus building have attracted significant attention not only by researchers, but also in many real-life application areas (such as engineering, medicine, social sciences and so on), due to the increasing need for substituting individual decision making processes in favor of highly accepted group decisions. Given the importance of group decision making and supporting consensual decisions across numerous areas, different researchers have proposed in the literature a variety of models and approaches to support high-quality collective decisions. These approaches have classically been limited to dealing with a low number of experts. However, nowadays environments and technological advances increasingly enable the participation of large groups in decision processes. This implies that LGDM problems, where a larger and more diverse group of participants take part, are attaining more interest in the last years. LGDM problems lay out new difficulties and challenges that most of the existing small group-focused approaches can no longer deal with properly:

- **Developing scalable and distributed decision support architectures**, capable of fluently managing large amounts of group and/or decision criteria information, as well as computational and communication processes; and accommodate highly distributed decision making where participants may be geographically separated.
- **Identifying and managing participants' behaviors**, particularly patterns of non-cooperative behavior exhibited by individuals who might attempt to deviate the decision outcomes in their favor.

I. Palomares Carrascosa, *Large Group Decision Making*, SpringerBriefs in Computer Science, https://doi.org/10.1007/978-3-030-01027-0_6

- **Defining cost-effective approaches for LGDM and building consensus**, by reducing the effort and temporal cost invested by experts in making such consensual decisions.
- **Monitoring the progress of the LGDM process** with the aid of data visualization or similar monitoring tools that allow analysts to gain a rapid and clear insight about the status of the problem being undertaken.
- **Implementing adequate strategies for weighting and aggregating decision information** in situations where multiple forms of uncertainty, different sources of information, social relationships or other inter-dependencies among participants arise.
- **Incorporating additional knowledge and data sources** stemming e.g. from the social media, relationships between participants, previous participation history, etc., that might be deemed as relevant to the decision problem at hand.

The above challenges have fueled a strong shift towards LGDM research in the last years, i.e. decision making and consensus reaching studies specifically focused on handling large-group decisions effectively and overcoming some of these challenges, often in a specific application domain. This book reviewed—to the author's knowledge—most of the extant literature in this topic, pointing out (1) the methodologies and specific considerations to be made for coping with large group decision situations, (2) showing the advantages and added value of each surveyed methodology, and (3) highlighting some examples of applications and implementations of extant approaches into LGDSS. Some representative examples of state-of-the-art LGDM solutions, related to the author's recent research, include:

- A multi-agent based consensus support system, based on an agent-driven architecture that allows a high level of scalability to support large groups efficiently. The system incorporates a novel agent-based semi-supervised autonomy approach, aimed at minimizing the human supervision of experts to revise and modify their preferences throughout a CRP [104].
- An extended consensus model with a methodology to manage non-cooperative behaviors in consensus reaching processes, that in conjunction with fuzzy clustering techniques, facilitates the detection and effective management of non cooperating subgroups and individuals [105]. Follow-up models to deal with non-cooperative behaviors integrate the analysis of both non-cooperation and positive cooperation patterns shown by individuals, for a more comprehensive assessment of their overall behavior [130].
- A graphical monitoring tool of preferences that facilitates a visual analysis of non-cooperative behaviors, minority and majority opinions, the evolution of preferences across a CRP, and other aspects of interest in LGDM processes where the analytical (non-visual) monitoring of the existing information becomes impracticable [107].
- A consensus model for MCLGDM that integrates social trust information between participants and uses it along with non-cooperative behavior detection to accurately assign importance weights to the opinions of participants [192].

Chapters 4 and 5 provided an exhaustive literature review comprising over 70 LGDM publications, which encompass the remarkable efforts made by the scientific community, classified into the following six research trends.

- *Subgroup Clustering.*
- *LGDM Methods.*
- *Consensus Approaches for LGDM.*
- *Behavior management and modeling.*
- *Theory and Interdisciplinary Studies.*
- *Large Group Decision Support Systems (LGDSS).*

For each trend, its associated contributions were discussed under several research themes and specialties which the different researchers focused on.

6.2   Lessons Learnt and Future Research Directions

It must be noted that LGDM is still a young and not solidly consolidated sub-area of group, multi-criteria and consensus decision making under uncertainty. However, it is rapidly gaining attention and benefiting from numerous advances, as reported in the literature review statistics (see Chap. 4) that show the increasing trend in the number of published LGDM contributions per annum, thereby justifying its growing importance in the years to come. This is probably the main lesson learned during the elaboration of this book, therefore it became necessary to provide the LGDM community with a first point of reference on the research topic at this stage, for both interested and acquainted readers. That is, essentially, the primary purpose of the book.

There are still a large number of pending challenges and shortages in the research field of GDM and consensus, particularly within LGDM. Some of these challenges have been previously identified throughout this book, and they are summarized below with the aim of establishing a series of guidelines for future research.

- Incorporating principles from psychology, e.g. behavioral and cognitive sciences, into LGDM research, so as to gain a better understanding of the motivations of participants' behavior and attitudes in these decisions processes, particularly when they involve reaching consensus before making a collective decision.
- Putting a greater focus on real deployed implementations of LGDM models, for instance into domain-independent Web LGDSS platforms or mobile apps that facilitate decision making processes across large groups situated anywhere. Integrating such next-generation LGDSS with "linked" sources of relevant partic- ipants' information, e.g. social media profiles, contextual decision information, etc., is another interesting direction for future research. As demonstrated in Chap. 5, very few models and methods have been deployed into real system implementations, thus this is a particular important gap to be bridged in the field.

- Definition of an established set of metrics for evaluating the performance of CRPs in LGDM, into a framework that enables the comparison between different proposals of consensus models.
- Exploring novel collective decision making paradigms including recommender systems for large groups, or autonomous intelligent systems undertaking collaborative decision making processes with each other or together with human participants.

Correction to: Large Group Decision Making

Correction to:
I. Palomares Carrascosa, *Large Group Decision Making*,
SpringerBriefs in Computer Science,
https://doi.org/10.1007/978-3-030-01027-0

This book was inadvertently published with wrong author names in Chaps. 2 and 4. Iván Palomares Carrascosa has now been rightly addressed as the author in these chapters and the research scholars have been acknowledged in the front matter.

The updated online version of this book can be found at
https://doi.org/10.1007/978-3-030-01027-0_2
https://doi.org/10.1007/978-3-030-01027-0_4
https://doi.org/10.1007/978-3-030-01027-0

References

1. Alcantud, J.C.R., de Andrés, R.: A fuzzy viewpoint of consensus measures in social choice. ESTYLF 2014 Proceedings: XVII Spanish Conference on Fuzzy Logic and Technologies, pp. 87–92, 2014.
2. Alonso, S., Pérez, I.J., Cabrerizo, F.J., Herrera-Viedma, E.: A Fuzzy Group Decision Making Model for Large Groups of Individuals. In: Proceedings of FUZZ-IEEE 2009, pp. 643–648, 2009.
3. Arrow, K.J.: A difficulty in the concept of social welfare. Journal of Political Economy, 58(4), pp. 328–346, 1950.
4. Atanassov, K.T.: Intuitionistic Fuzzy Sets. Fuzzy Sets and Systems, 20(1), pp. 87–96, 1986.
5. B´ack, E., Esaiasson, P., Gilljam, M., Svenson, O., Lindholm, T.: Post-Decision Consolidation in Large Group decision-making. Cognition and Neurosciences. Scandinavian Journal of Psychology, 52, pp. 320–328, 2011.
6. Baker, K.R.: Management Science: An Introduction to the Use of Decision Models. Wiley (NY), 1985.
7. M. Behzadian, S.K. Otaghsara, M. Yazdani, J. Ignatius.: A state-of the-art survey of TOPSIS applications. Expert Systems with Applications, 39, pp. 13051–13069, 2012.
8. Beliakov, G., Pradera, A., Calvo, T.: Aggregation functions: a guide for practitioners. Springer Studies in Fuzziness and Soft Computing (reprint), Springer, 2010.
9. Bellman, R.E., Zadeh, L.A.: Decision-making in a fuzzy environment. Management Science, 17(4), pp. 141–164, 1970.
10. Ben-Arieh, D., Chen. Z.: Linguistic labels aggregation and consensus measure for autocratic decision-making using group recommendations. IEEE Transactions on Systems, Man and Cybernetics, Part A: Systems and Humans, 36(1), pp. 558–568, 2006.
11. Ben-Arieh, D., Chen, Z.: Linguistic group decision-making: opinion aggregation and measures of consensus. Fuzzy Optimization and Decision Making, 5(4), pp. 371–386, 2006.
12. Blondel, V.D., Guillaume, J.L., Lambiotte, R., Lefebvre, E.: Fast unfolding of communities in large networks. Journal of Statistical Mechanics: Theory and Experiment, 2008 (10), pp. 1–12, 2008.
13. Bolloju, N.: Aggregation of analytic hierarchy process models based on similarities in decision makers' preferences. European Journal of Operational Research, 128, pp. 499–508, 2001.
14. Bordogna, G., Fedrizzi, M., Pasi, G.: A linguistic modeling of consensus in group decision making based on OWA operators. IEEE Transactions on Systems, Man and Cybernetics - Part A: Systems and Humans, 27(1), pp. 126–133, 1997.

I. Palomares Carrascosa, *Large Group Decision Making*, SpringerBriefs in Computer Science, https://doi.org/10.1007/978-3-030-01027-0

15. Bryson, N.: Group decision-making and the analytic hierarchy process. exploring the consensus-relevant information content. Computers and Operations Research, 23(1), pp. 27–35, 1996.
16. Bouyssou, D., Dubois, D., Prade, H., Pirlot, M. (Eds.): Decision making process: concepts and methods. Wiley-ISTE, 2009.
17. Bullock, S., Crowder, R., Pitonakova, L.: Task allocation in foraging robot swarms: The role of information sharing. Proceedings of the European Conference on Artificial Life 13, pp. 306–313, 2016.
18. Butler, C.T.L., Rothstein, A.: On conflict and consensus: A handbook on formal consensus decision making. Food Not Bombs Publishing (Takoma Park), 2006.
19. Cai, C.-G., Xu, X.-H., Wang, P., Chen, X.-H.: A multi-stage conflict style large group emergency decision-making method. Soft Computing, 21, pp. 5765–5778, 2017.
20. Campanella, G., Ribeiro, R.: A framework for dynamic multiple-criteria decision making. Decision Support Systems, 52, pp. 52–60, 2011.
21. Carlsson, C., Ehrenberg, D., Eklund, P., Fedrizzi, M., Gustafsson, P., Lindholm, P., Merkuryeva, G., Riissanen, T., Ventre, A.G.S.: Consensus in distributed soft environments. European Journal of Operational Research, 61(1–2), pp. 165–185, 1992.
22. Carneiro, J., Saraiva, P., Martinho, D., Marreiros, G., Novais, P.: Representing decision-makers using styles of behavior: an approach designed for group decision support systems. Cognitive Systems Research, 47, pp. 109–132, 2018.
23. Cartlidge, J., Cliff, D.: Modelling complex financial markets using real-time human-agent trading experiments. In Chen S.H. et al. (Eds.): Complex Systems Modeling and Simulation in Economics and Finance, Springer, 2018.
24. Carvalho, G., Vivacqua, A.S., Souza, J.M., Medeiros, S.P.J.: LaSca: a Large Scale Group Decision Support System. Proceedings of 12th International Conference on Computer Supported Cooperative Work in Design. Xi'an (China), 2008.
25. Chadwick, A.: Web 2.0: New challenges for the study of e-democracy in an era of informational exuberance. I/S: A Journal of Law and Policy for the Information Society, 5(1), pp. 9–41, 2009.
26. Chavez, A., Maes, P.: Kasbah: An agent marketplace for buying and selling goods. Procs. 1st International Conference on the Practical Application of Intelligent Agents and Multi-Agent Technology, pp. 75–90, 1996.
27. Chen, J.L., Chen, C., Wang, C.C., Jiang, X.: Measuring soft consensus in uncertain linguistic group decision-making based on deviation and overlap degrees. International Journal of Innovative Management, Information & Production, 2(3), pp. 25–33, 2011.
28. Chiclana, F., Herrera-Viedma, E., Alonso, S., Marques-Pereira, R.: Preferences and consistency issues in group decision making. In H. Bustince et al. (Eds.): Fuzzy Sets and Their Extensions: Representation, Aggregation and Models. Intelligent Systems from Decision Making to Data Mining, Web Intelligence and Computer Vision. Studies in Fuzziness and Soft Computing, 220, pp. 219–237, Springer Verlag, 2008.
29. Chin, K.S., Xu, D.L., Yang, J.B., Lam, J.P.-K.: Group-based ER-AHP system for product project screening. Expert Systems with Applications, 35(4), pp. 1909–1929, 2008.
30. Choo, E.U., William C.W.: A common framework for deriving preference values from pairwise comparison matrices. Computers & Operations Research, 31(6), pp. 893–908, 2004.
31. Choudhury, A.K., Shankar, R., Tiwari, M.K.: Consensus-based intelligent group decision-making model for the selection of advanced technology. Decision Support Systems, 42(3), pp. 1776–1799, 2006.
32. Cole, J.D., Sage, A.P.: Multi-person decision analysis in large scale systems - group decision making. Journal of the Franklin Institute, Apr 1975, pp. 245–268.
33. Crosscombe, M., Lawry, J.: Exploiting vagueness for multi-agent consensus. Multi-agent and Complex Systems, Studies in Computational Intelligence, vol. 670, pp. 67–78, Springer, 2017.

34. Dong, Y., Wu, Y., Zhang, H., Zhang, G.: Multi-granular unbalanced linguistic distribution assessments with interval symbolic proportions. Knowledge-based Systems, 82, pp. 139–151, 2015.
35. Dong, Y., Xu, J.: Consensus building in group decision making - searching the consensus path with minimum adjustments. Springer, 2016.
36. Dong, Y., Zhang, H., Herrera-Viedma, E.: Integrating experts' weights generated dynamically into the consensus reaching process and its applications in managing non-cooperative behaviors. Decision Support Systems, 84, pp. 1–15, 2016.
37. Dong, Y., Ding, Z., Martinez, L., Herrera, F.: Managing consensus based on leadership in opinion dynamics. Information Sciences, 397–298, pp. 187–205, 2016.
38. Dong, Y., Zhao, S., Zhang, H., Chiclana, F., Herrera-Viedma, E.: A self-management mechanism for non-cooperative behaviors in large-scale group consensus reaching processes. IEEE Transactions on Fuzzy Systems, In press. https://doi.org/10.1109/TFUZZ.2018.2818078.
39. Dong, Y., Zha, Q., Zhang, H., Kou, G., Fujita, H., Chiclana, F., Herrera-Viedma, E.: Consensus Reaching in Social Network Group Decision Making: Research Paradigms and Challenges. Knowledge-based Systems, In press. https://doi.org/10.1016/j.knosys.2018.06.036.
40. Dong, Y., Zhan, M., Kou, G., Ding, Z., Liang, H.: A survey of the fusion process in opinion dynamics. Information Fusion, 43, pp. 57–65, 2018.
41. Doumpos, M., Zopounidis, C.: Multicriteria Decision Aid Classification Methods. Springer Science & Business Media, 2006.
42. Dymova, L, Savastjanov, P., Tikhonenko, A.: A direct interval extension of TOPSIS method. Expert Systems with Applications, 40, pp. 4841–4847, 2013.
43. Felfernig, A., Boratto, L., Stettinger, M., Tkalcic, M.: Group Recommender Systems - an Introduction. SpringerBriefs in Electrical and Computer Engineering, Springer, 2018.
44. Flach, P.: Machine Learning: The Art and Science of Algorithms that make sense of Data. Cambridge University Press, 2012.
45. French, J.R., John, R.P.: A formal theory of social power. Psychological Review, 63(3), pp. 181–194, 1956.
46. García-Lapresta, J.L., Llamazares, B.: Aggregation of fuzzy preferences: some rules of the mean. Social Choice and Welfare, 17(4), pp. 673–690.
47. García-Alcaraz, J.L., Martínez-Loya, V., Díaz-Reza, R., Avelar, L., Canales, I.: A multicriteria decision support system framework for computer selection. In R. Valencia-García et al. (Eds.): Exploring Intelligent Decision Support Systems, Studies in Computational Intelligence, 764, pp. 89–110, 2018.
48. Goel, A., Lee, D.T.: Towards large-scale deliberative decision-making: small groups and the importance of triads. EC '16 Proceedings of the 2016 ACM Conference on Economics and Computation, pp. 287–303, 2016.
49. Gomes, L.F.A.M., Lima, M.P.P.: TODIM: Basic and application to multicriteria ranking of projects with environmental impacts. Foundations of Computing and Decision Sciences, 16(3), pp. 113–127, 1991.
50. Gong, Z.W., Forrest, J., Yang, Y.J.: The optimal group consensus models for 2-tuple linguistic preference relations. Knowledge-based systems, 37, pp. 427–437, 2013.
51. González-Artega, T., de Andrés, R., Chiclana, F.: A new measure of consensus with reciprocal preference relations. Knowledge-based Systems, 107(C), pp. 104–116, 2016.
52. Gou, X., Xu, Z., Herrera, F.: Consensus Reaching Process for Large-scale Group Decision Making with Double Hierarchy Hesitant Fuzzy Linguistic Preference Relations. Knowledge-based Systems, 157, pp. 20–33, 2018.
53. Gupta, M.: Consensus building process in group decision making - an adaptive procedure based on group dynamics. IEEE Transactions on Fuzzy Systems, In press. https://doi.org/10.1109/TFUZZ.2017.2755581
54. Hansson, S.O.: Decision Theory: A brief introduction. Royal Institute of Technology (KTH), Stockholm, 2005.

55. Herrera, F., Herrera-Viedma, E., Verdegay, J.L.: Direct approach processes in group decision making using linguistic OWA operators. Fuzzy Sets and Systems, 79(2), pp. 175–190, 1996.

56. Herrera, F., Herrera-Viedma, E., Verdegay, J.L.: A model of consensus in group decision making under linguistic assessments. Fuzzy Sets and Systems, 78(1), pp. 73–87, 1996.

57. Herrera, F., Herrera-Viedma, E., Verdegay, J.L.: Linguistic measures based on fuzzy coincidence for reaching consensus in group decision making. International Journal of Approximate Reasoning, 16(3–4), pp. 309–334, 1997.

58. Herrera, F., Herrera-Viedma, E., Verdegay, J.L.: A rational consensus model in group decision making using linguistic assessments. Fuzzy Sets and Systems, 88(1), pp. 31–49, 1997.

59. Herrera-Viedma, E., Herrera, F., Chiclana, F.: A consensus model for multiperson decision making with different preference structures. IEEE Transactions on Systems, Man and Cybernetics, Part A: Systems and Humans, 32(3), pp. 394–402, 2002.

60. Herrera-Viedma, E., Alonso, S., Chiclana, F., Herrera, F.: A consensus model for group decision making with incomplete fuzzy preference relations. IEEE Transactions on Fuzzy Systems, 15(5), pp. 863–877, 2007.

61. Herrera-Viedma, E., García-Lapresta, J.L., Kacprzyk, J., Fedrizzi, M., Nurmi, H., Zadrozny, S. (Eds.): Consensual Processes. Studies in Fuzziness and Soft Computing, 267, Springer, 2011.

62. Hirano, K.: Decision Theory in Econometrics. In: Durlauf S.N., Blume L.E. (eds) Microeconometrics. The New Palgrave Economics Collection. Palgrave Macmillan, London, 2010.

63. Hoegen, A., Steininger, D., Veit, D.: How do investors decide? An interdisciplinary review of decision-making in crowdfunding. Electronic Markets, Oct. 2017, pp. 1–27, 2017.

64. Husain, A.J.A.: A multi-agent system for scalable group decision making.

65. Hwang, C. L., Yoon, K. P.: Multiple attribute decision making: Methods and applications. New York, Springer-Verlag, 1981.

66. Ishizaka, A., Markus L.: How to derive priorities in AHP: a comparative study. Central European Journal of Operations Research, 14(4), pp. 387–400, 2006.

67. Kacprzyk, J.: Group decision making with a fuzzy linguistic majority. Fuzzy Sets and Systems, 18(2), pp. 105–118, 1986.

68. Kacprzyk, J.: On some fuzzy cores and 'soft' consensus measures in group decision making. In J. Bezdek (Ed.), The Analysis of Fuzzy Information, pp. 119–130, 1987.

69. Kacprzyk, J., Fedrizzi, M.: A "soft" measure of consensus in the setting of partial (fuzzy) preferences. European Journal of Operational Research, 34(1), pp. 316–325, 1988.

70. Kacprzyk, J., Fedrizzi, M., Nurmi, H.: Group decision making and consensus under fuzzy preferences and fuzzy majority. Fuzzy Sets and Systems, 49(1), pp. 21–31, 1992.

71. Kacprzyk, J., Zadrozny, S.: Soft computing and web intelligence for supporting consensus reaching. Soft Computing, 14(8), pp. 833–846, 2010.

72. Kline, J.A.: Orientation and group consensus. Central States Speech Journal, 23, pp. 44–47, 1972.

73. Khorshid, S.: Soft consensus model based on coincidence between positive and negative ideal degrees of agreement under a group decision-making fuzzy environment. Experts systems with applications, 37(5), pp. 3977–3985, 2010.

74. Kohonen, T.: Self-organizing maps. Heidelberg, Springer, 1995.

75. Labella, A., Liu, Y., Rodríguez, R.M., Martínez, L.: Analyzing the performance of classical consensus models in large scale group decision making: A comparative study. Applied Soft Computing, 67, pp. 677–690, 2018.

76. Lai, S.K.: A preference-based interpretation of AHP. Omega, 23, pp. 453–462, 1995.

77. Lawry, J., Tang, Y.: Uncertainty modelling for vague concepts: A prototype theory approach. Artificial Intelligence, 173(18), pp. 1539–1558, 2009.

78. Li, Y., Lian, X., Lu, C., Wang, Z.: A large group decision making approach based on TOPSIS framework with unknown weights information. MATEC Web of Conferences (13th Global Congress on Manufacturing and Management, GCMM 2016), vol. 100, 2017.

79. Lin, C.-C.: A revised framework for deriving preference values from pairwise comparison matrices. European Journal of Operational Research, 176 (2), pp. 1145–1150, 2007.

80. Liu, H.C., You, X.Y., Tsung, F., Ji, P.: An improved approach for failure mode and effect analysis involving large group of experts: an application to the healthcare field. Quality Engineering, In press. https://doi.org/10.1080/08982112.2018.1448089.
81. Liu, B., Shen, Y., Chen, X., Sun, H., Chen, Y.: A complex multi-attribute large-group PLS decision-making method in the interval-valued intuitionistic fuzzy environment. Applied Mathematical Modelling, 38, pp. 4512–4527, 2014.
82. Liu, B., Huo, T., Liao, P., Gong, J., Xue, B.: A group decision-making aggregation model for contractor selection in large scale construction projects based on two-stage partial least squares (PLS) path modeling. Group Decision and Negotiation, 24(5), pp. 855–883, 2014.
83. Liu, B., Shen, Y., Chen, X., Che, Y., Wang, X.: A partial binary tree DEA-DA cyclic classification model for decision makers in complex multi-attribute large-group interval-valued intuitionistic fuzzy decision-making problems. Information Fusion, 18, pp. 119–130, 2014.
84. Liu, B., Chen, Y., Shen, Y., Sun, H., Xu, X.: A complex multi-attribute large-group decision making method based on the interval-valued intuitionistic fuzzy principal component analysis model. Soft Computing, 18, pp. 2149–2160, 2014.
85. Liu, B., Shen, Y., Zhang, W., Chen, X., Wang, X.: An interval-valued intuitionistic fuzzy principal component analysis model-based method for complex multi-attribute large-group decision-making. European Journal of Operational Research, 245, pp. 209–225, 2015.
86. Liu, B., Shen, Y., Chen, Y., Chen, X., Wang, Y.: A two-layer weight determination method for complex multi-attribute large-group decision-making experts in a linguistic environment. Information Fusion, 23, pp. 156–165, 2015.
87. Liu, Y., Fan, Z.P., Zhang, X.: A method for large group decision-making based on evaluation information provided by participators from multiple groups. Information Fusion, 29, pp. 132–141, 2016.
88. Liu, Y., Liang, C., Chiclana, F., Wu, J.: A trust induced recommendation mechanism for reaching consensus in group decision making. Knowledge-based Systems, 119, pp. 221–231, 2017.
89. Liu, B., Guo, S., Yan, K., Li, L., Wang, X.: Double weight determination method for experts of complex multi-attribute large-group decision-making in interval-valued intuitionistic fuzzy environment. Journal of Systems Engineering and Electronics, 28(1), pp. 88–96, 2017.
90. Muller, M.J.: Participatory Design: The Third Space in HCI. IBM Technical Report #01-04, 2002.
91. Xiang, L.: Energy network dispatch optimization under emergency of local energy shortage with web tool for automatic large group decision-making. Energy, 120, pp. 740–750, 2017.
92. Lootsma, F.A.: Scale sensitivity in the multiplicative AHP and SMART. Journal of Multi-Criteria Decision Analysis, 2(2), pp. 87–110, 1993.
93. Lu, J., Zhang, G., Ruan, D., Wu, F.: Multi-objective group decision making - methods, software and applications with fuzzy set techniques. World Scientific Series in Electrical and Computer Engineering, 6, 2007.
94. Martinez, L., Montero, J.: Challenges for improving consensus reaching process in collective decisions. New Mathematics and Natural Computation, 3(2), pp. 203–217, 2007.
95. Martinez, L, Herrera, F.: An overview on the 2-tuple linguistic model for computing with words in decision making: Extensions, applications and challenges. Information Sciences, 207, pp. 1–18, 2012.
96. Mata, F., Martinez, L., Herrera-Viedma, E.: An adaptive consensus support model for group decision-making problems in a multigranular fuzzy linguistic context. IEEE Transactions on Fuzzy Systems, 17(2), pp. 279–290, 2009.
97. Mendel, J., John, R.I.B.: Type-2 Fuzzy Sets made Simple. IEEE Transactions on Fuzzy Systems, 10(2), pp. 117–127, 2002.
98. Nyerges, T., Aguirre, R.W.: Public Participation in Analytic-Deliberative Decision Making: Evaluating a Large-Group Online Field Experiment. Annals of the Association of American Geographers, 101(3), pp. 561–586, 2011.

99. Orlovsky, S.: Decision-making with a fuzzy preference relation. Fuzzy Sets and Systems, 1(3), 155–167.

100. Palomares, I., Sánchez, P., Quesada, F., Mata, F., Martínez, L.: COMAS - A multi-agent system for performing consensus processes. In Abraham, A. et al. (Eds.): Procs. International Symposium on Distributed Computing and Artificial Intelligence (DCAI 2011). Advances in Intelligent and Soft Computing, 91, pp. 125–132, Springer, 2011.

101. Palomares, I., Liu, J., Xu, Y., Martínez, L.: Modelling experts' attitudes in group decision making. Soft Computing, 16(10), pp. 1755–1766, 2012.

102. Palomares, I., Quesada, F., Martinez, L.: Multi-agent-based semi-supervised consensus support system for large-scale group decision making. In Z. Wen and T. Li (eds.): Foundations of Intelligent Systems (ISKE 2013 Proceedings), Advances in Intelligent Systems and Computing 277, pp. 241–251, Springer, 2014.

103. Palomares, I., Estrella, F.J., Martinez, L., Herrera, F.: Consensus under a fuzzy context - taxonomy, analysis framework AFRYCA and experimental case of study. Information fusion, 20, pp. 252–271, 2014.

104. Palomares, I., Martinez, L.: A semisupervised multiagent system model to support consensus-reaching processes. IEEE Transactions on Fuzzy Systems, 22(4), pp. 762–777, 2014.

105. Palomares, I., Martínez, L., Herrera, F.: A consensus model to detect and manage non-cooperative behaviors in large-scale group decision making. IEEE Transactions on Fuzzy Systems, 22(3), pp. 516–530, 2014.

106. Palomares, I., Quesada, F., Martínez, L.: An approach based on computing with words to manage experts behavior in consensus reaching processes with large groups. Procs. 2014 IEEE International Conference on Fuzzy Systems (FUZZ-IEEE), 2014.

107. Palomares, I., Martínez, L., Herrera, F.: MENTOR: A graphical monitoring tool of preferences evolution in large-scale group decision making. Knowledge-based Systems, 58 (Spec.Iss.), pp. 66–74, 2014.

108. Palomares, I.: Multi-agent System to model consensus processes in large-scale group decision making using soft computing techniques. PhD Thesis, University of Jaén (Spain), 2014.

109. Quesada, F., Palomares, I., Martínez, L.: Managing experts behavior in large-scale consensus reaching processes with uninorm aggregation operators. Applied Soft Computing, 35, pp. 873–887, 2015.

110. Palomares, I., Killough, R., Bauters, K., Liu, W., Hong, J.: A collaborative multiagent framework based on online risk-aware planning and decision-making. In Proceedings of ICTAI'16 International Conference, 2016.

111. Palomares, I., Sellak, H., Ouhbi, B., Frikh, B.: Adaptive, Semi-Supervised Consensus Model for Multi-Criteria Large Group Decision Making in a Linguistic Setting. In ISKE 2017 Proceedings: 12th International Conference on Intelligent Systems and Knowledge Engineering, 2017.

112. Palomares, I., Crosscombe, M., Chen, Z.S., Lawry, J.: Dual Consensus Measure for Multi-Perspective Multi-Criteria Group Decision Making. Proceedings of IEEE International Conference on Systems, Man and Cybernetics, IEEE SMC'18. 2018.

113. Parreiras, R.O., Ekel, P., Bernardes, A.: A dynamic consensus scheme based on a nonreciprocal fuzzy preference relation modeling. Information Sciences, 211(1), pp. 1–17, 2012.

114. Parreiras, R.O., Ekel, P., Bernardes, A.: A dynamic consensus scheme based on a nonreciprocal fuzzy preference relation modeling. Information Sciences, 211(1), pp. 1–17, 2012.

115. Peterson, M.: An Introduction to Decision Theory (Cambridge Introductions to Philosophy). Cambridge University Press, 2011.

116. Rodríguez, M.A.: Advances towards a general-purpose societal-scale human-collective problem-solving engine. Procs. 2004 IEEE International Conference on Systems, Man and Cybernetics (SMC), pp. 206–211, 2004.

117. Rodriguez, M.A.: Social decision making with multi-relational networks and grammar-based particle swarms. Procs. 40th Hawaii International Conference on System Sciences, 2007.

118. Roubens, M.: Fuzzy sets and decision analysis. Fuzzy Sets and Systems, 90(2), pp. 199–206, 1997.

119. Russell, S., Norvig, P.: Artificial Intelligence: A Modern Approach (3rd Ed.). Pearson, 2016.
120. Russo, R.D.F.S.M., Camanho, R.: Criteria in AHP: A systematic review of literature. Procedia Computer Science, pp. 1123–1132, Elsevier, 2015.
121. Saaty, T.L.: Highlights and critical points in the theory and application of the analytic hierarchy process. European Journal of Operational Research, 74, pp. 426–447, 1994.
122. Saaty, T.L.: Decision-making with the AHP: why is the principal eigenvector necessary. European Journal of Operational Research, 145, pp. 85–91, 2003.
123. Saaty, Thomas L. "A scaling method for priorities in hierarchical structures." Journal of mathematical psychology 15.3 (1977): 234–281.
124. Saaty, Thomas L. "The Analytical Hierarchy Process, Planning, Priority." Resource Allocation. RWS Publications, USA (1980).
125. Saaty, Thomas L. "Decision making with the analytic hierarchy process." International journal of services sciences 1.1 (2008): 83–98.
126. Saint, S., Lawson, J.R.: Rules for reaching consensus. A modern approach to decision making. Jossey-Bass, 1994.
127. Spillman, B., Bezdek, J., Spillman, R.: Development of an instrument for the dynamic measurement of consensus. Communication Monographs, 46(1), pp. 1–12, 1979.
128. Squillante, M.: Decision making in social networks. International Journal of Intelligent Systems, 25(3), Special Issue, 2010.
129. Shannon, C.E.: A mathematical theory of communication. The Bell System Technical Journal, 27, pp. 379–423, 623–656, 1948.
130. Shi, Z.J., Wang, X.Q., Palomares, I., Guo, S.J., Ding, R.X.: A novel consensus model for multi-attribute large-scale group decision making based on comprehensive behavior Classification and adaptive weight updating. Knowledge-based Systems, In Press. https://doi.org/10.1016/j.knosys.2018.06.002
131. Shum, S., Cannavacciuolo, L., De Liddo, A., Iandoli, L., Quinto, I.: Using social network analysis to support collective decision-making processes. International Journal of Decision Support System Technology, 3(2), pp. 15–31, 2011.
132. Smith, J.E., Winterfeldt, D.: Decision Analysis in "Management Science". Management Science, 50(5), pp. 561–574, 2004.
133. Sorin, N., Dzitac, S., Dzitax, I.: Fuzzy TOPSIS: a general view. Procedia Computer Science, pp. 823–831, Elsevier, 2016.
134. Soto, R., Robles-Baldenegro, M.E., López, V.: MQDM: An iterative fuzzy method for group decision making in structured social networks. International Journal of Intelligent Systems, 32, pp. 17–30, 2017.
135. Srdjevic, B.: Linking analytic hierarchy process and social choice methods to support group decision-making in water management. Decision Support Systems, 42, pp. 2261–2273, 2007.
136. Stamatis, .D.H.: Failure mode and effect analysis: FMEA from theory to execution (2nd Ed). American Society for Quality Press, 2003.
137. Tanino, T.: Fuzzy preference orderings in group decision making. Fuzzy Sets and Systems, 12(2), pp. 117–131, 1984.
138. Tapia-Rosero, A., De Tré, G.: Evaluating relevant opinions within a large group. Procs. International Conference on Fuzzy Computation Theory and Applications (FCTA-2014), pp. 76–86, 2014.
139. Tapia-Rosero, A., Bronselaer, A., De Tré, G.: A method based on shape-similarity for detecting similar opinions in group decision-making. Information Sciences, 258, pp. 291–311, 2014.
140. Tapia-Rosero, A., De Mol, R., De Tré, G.: Handling uncertainty degrees in the evaluation of relevant opinions within a large group. In J.J. Merelo et al. (Eds.), Computational Intelligence, Studies in Computational Intelligence, 620, Springer, 283–299, 2015.
141. Tapia-Rosero, A.: Handling a Large Number of Preferences in a Multi-Level Decision-Making Process. PhD Thesis, Ghent University (Belgium), 2016.
142. Tapia-Rosero, A., Bronselaer, A., De Mol, R., De Tré, G.: Fusion of preferences from different perspectives in a decision-making context. Information fusion, 29, pp. 120–131, 2016.

143. Tocqueville, A.: Democracy in America (2nd Ed.). Saunders and Otley (London), 1840.
144. Torra, V.: Hesitant Fuzzy Sets. International Journal of Intelligent Systems, 25(6), pp. 529–539, 2010.
145. Torra, V., Mesiar, R., Baets, B.: Aggregation functions in theory and practice. Springer Advances in Intelligent Systems and Computing (Proceedings AGOP 2017), Springer, 2017.
146. Turoff, M., Hiltz, S.R., Cho, H.-K., Li, Z., Wang, Y.: Social Decision Support Systems (SDSS). Procs. 35th Hawaii International Conference on System Sciences, 2002.
147. Ureña, R., Chiclana, F., Morente-Molinera, J.A., Herrera-Viedma, E.: Managing incomplete preference relations in decision making: A review and future trends. Information Sciences 302, pp. 14–32, 2015.
148. von Neumann, J., Morgenstern, O.: Theory of Games and Economic Behavior. Princeton University Press (NJ), 1944.
149. Wang, J.Q.: Multi-criteria large-group linguistic decision-making approach with incomplete certain information. Procs. Chinese Control and Decision Conference, 2009. CCDC '09, 2009.
150. Wu, Z., Xu, J.: A consistency and consensus based decision support model for group decision making with multiplicative preference relations. Decision Support Systems, 52(3), pp. 757–767, 2012.
151. Wu, J., Chiclana, F., Herrera-Viedma, E.: Trust based consensus model for social network in an incomplete linguistic information context. Applied Soft Computing, 35, pp. 827–839, 2015.
152. Wu, T., Liu, X.W.: An interval type-2 fuzzy clustering solution for large-scale multiple-criteria group decision-making problems. Knowledge-based Systems, 144, pp. 118–127, 2016.
153. Wu, T., Liu, X., Qin, J.: A linguistic solution for double large-scale group decision-making in E-commerce. Computers & Industrial Engineering, 116, pp. 97–112, 2018.
154. Wu, T., Liu, X., Liu, F.: An interval type-2 fuzzy TOPSIS model for large scale group decision making problems with social network information. Information Sciences, 42, pp. 392–410, 2018.
155. Wu, Z., Xu, J.: A consensus model for large-scale group decision making with hesitant fuzzy information and changeable clusters. Information Fusion, 41, pp. 217–231, 2018.
156. Xia, M., Xu, Z., Chen, J.: Algorithms for improving consistency or consensus of reciprocal [0,1]-valued preference relations. Fuzzy sets and systems, 216(Spec. Iss.), pp. 108–133, 2013.
157. Xia, M., Xu, Z., Chen, N.: Some Hesitant Fuzzy Aggregation Operators with Their Application in Group Decision Making. Group Decision and Negotiation. 22(2), pp. 259–279, 2013.
158. Xiang, L.: Energy network dispatch optimization under emergency of local energy shortage with web tool for automatic large group decision-making. Energy, 120, pp. 740–750, 2017.
159. Xu, X.H., Chen, X., Wang, H.: A kind or large group decision making method on the utility value preference information of decision member. Procs. 4th International Conference on Wireless Communications, Networking and Mobile Computing, 2008. WiCOM '08, 2008.
160. Xu, Z.: An automatic approach to reaching consensus in multiple attribute group decision making. Computers & Industrial Engineering, 56(4), pp. 1369–1374, 2009.
161. Xu, X.H., Ahn, J., Chen, X., Zhou, Y.: Conflict measure model for large group decision based on interval intuitionistic trapezoidal fuzzy number and its application. Journal of Systems Science and Systems Engineering, 22(4), pp. 487–498, 2013.
162. Xu, X.H., Liang, D., Chen, X., Zhou, Y.: A risk elimination coordination method for large group decision-making in natural disaster emergencies. Human and Ecological Risk Assessments: An International Journal, 21(5), pp. 1314–1325, 2014.
163. Xu, X.H., Cai, C., Chen, X., Zhou, Y.: A multi-attribute large group emergency decision making method based on group preference consistency of generalized interval-valued trapezoidal fuzzy numbers. Journal of Systems Science and Systems Engineering, 24(2), pp. 211–228, 2015.

164. Xu, X.H., Zhong, X.Y., Chen, X.H., Zhou, Y.J.: A dynamical consensus method based on exit-delegation mechanism for large group emergency decision making. Knowledge-based Systems, 86, pp. 237–249, 2015.
165. Xu, X.H., Du, Z.J., Chen, X.H.: Consensus model for multi-criteria large-group emergency decision making considering non-cooperative behaviors and minority opinions. Decision Support Systems, 79, pp. 150–160, 2015.
166. Xu, X.H., Wang, B., Zhou, Y.: A method based on trust model for large group decision-making with incomplete information. Journal of Intelligent & Fuzzy Systems, 30(6), pp. 3551–3565, 2016.
167. Xu, X.H., Sun, Q., Pan, B., Liu, B.: Two-layer weight large group decision-making method based on multi-granular attributes. Journal of Intelligent and Fuzzy Systems, 33, pp. 1797–1807, 2017.
168. Xu, Y., Wen, X., Zhang, W.: A two-stage consensus method for large-scale multi-attribute group decision making with an application to earthquake shelter selection. Computers & Industrial Engineering, 116, pp. 113–129, 2018.
169. Xue, B., Xu, H.: A Whole Life Cycle Group Decision-Making Framework for Sustainability Evaluation of Major Infrastructure Projects. In K.W. Chau et al. (Eds.): Procs. 21st International Symposium on Advancement of Construction Management and Real Estate, pp. 129–141, Springer, 2018.
170. Yager, R.: On ordered weighted averaging aggregation operators in multi-criteria decision making. IEEE Transactions on Systems, Man and Cybernetics. 18(1), pp. 183–190, 1988.
171. Yager, R., Rybalov, A.: Uninorm aggregation operators. Fuzzy Sets and Systems. 80, pp. 111–120, 1996.
172. Yager, R., Filev, D.: Induced Ordered Weighted Averaging Operators. IEEE Transactions on Systems, Man and Cybernetics, 29, pp. 141–150, 1999.
173. Yager, R.: Penalizing strategic preference manipulation in multi-agent decision making. IEEE Transactions on Fuzzy Systems, 9(3), pp. 393–403, 2001.
174. Yang, Y., Fu, C., Chen, Y.-W., Xu, D.-L., Yang, S.-L.: A belief rule based expert system for predicting consumer preference in new product development. Knowledge-based Systems, 94, pp. 105–113, 2016.
175. Yu, W., Zhang, Z., Zhong, Q.Y.: A TODIM-Based Approach to Large-Scale Group Decision Making with Multi-Granular Unbalanced Linguistic Information. Procs. 2017 IEEE International Conference on Fuzzy Systems (FUZZ-IEEE), 2017.
176. Zadeh, L.A.: Fuzzy sets. Information and Control, 8(3), pp. 338–353. 1965.
177. Zadeh, L.A.: The concept of a linguistic variable and its application to approximate reasoning - I. Information Sciences, 8(3), pp. 199–249, 1975.
178. Zadeh, L.A.: The concept of a linguistic variable and its application to approximate reasoning - II. Information Sciences, 8(4), pp. 301–357, 1975.
179. Zadeh, L.A.: The concept of a linguistic variable and its application to approximate reasoning - II. Information Sciences, 9(1), pp. 43–80, 1975.
180. Zadeh, L.A.: A computational approach to fuzzy quantifiers in natural languages. Computing and Mathematics with Applications, 9(1), pp. 149–184, 1983.
181. Zadeh, L.A.: Fuzzy logic = computing with words. IEEE Transactions on Fuzzy Systems, 4(2), pp. 103–111, 1996.
182. Zahir, S.: Clusters in a group: decision making in the vector space formulation of the analytics hierarchy process. European Journal of Operational Research, 112, pp. 620–634, 1999.
183. Zeleny, M.: A concept of compromise solutions and the method of the displaced ideal. Computers & Operations Research, 1(3–4), pp. 379–396, 1974.
184. Zhang, G., Dong, Y., Xu, Y., Li, H.: Minimum-cost consensus models under aggregation operators. IEEE Transactions on Systems, Man and Cybernetics - Part A: Systems and Humans, 41(6), pp. 1253–1261, 2011.
185. Zhang, F., Ignatius, J., Lim, C.P., Goh, M.: A two-stage dynamic group decision making method for processing ordinal information. Knowledge-based Systems, 70, pp. 189–202, 2014.

186. Zhang, G., Dong, Y., Xu, Y.: Consistency and consensus measures for linguistic preference relations based on distribution assessments. Information Fusion, 17, pp. 46–55, 2014.

187. Zhang, F., Ignatius, J., Zhao, Y., Lim, C.P., Ghasemi, M., Ng, P.S.: An improved consensus-based group decision making model with heterogeneous information. Applied Soft Computing, 35, pp. 850–863, 2015.

188. Zhang, X.: A Novel Probabilistic Linguistic Approach for Large-Scale Group Decision Making with Incomplete Weight Information. International Journal of Fuzzy Systems, Aug. 2017, pp. 1–12, 2017.

189. Zhang, Z., Guo, C., Martínez, L.: Managing multigranular linguistic distribution assessments in large-scale multiattribute group decision making. IEEE Transactions on Systems, Man and Cybernetics: Systems, 47(11), pp. 3063–3076, 2017.

190. Zhang, H., Dong, Y., Chen, X.: The 2-Rank Consensus Reaching Model in the Multigranular Linguistic Multiple-Attribute Group Decision-Making. IEEE Transactions on Systems, Man and Cybernetics: Systems, In Press. https://doi.org/10.1109/TSMC.2017.2694429

191. Zhang, H., Dong, Y., Herrera-Viedma, E.: Consensus building for the heterogeneous large-scale GDM with the individual concerns and satisfactions. IEEE Transactions on Fuzzy Systems, 26(2), pp. 884–898, 2018.

192. Zhang, H., Palomares, I., Dong, Y., Wang, W.: Managing non-cooperative behaviors in consensus-based multiple attribute group decision making: An approach based on social network analysis. Knowledge-based Systems, In press. https://doi.org/10.1016/j.knosys.2018.06.008

193. Zhu, W.D., Liu, F., Chen, Y.W., Yang, J.B., Xu, D.L., Wang, D.P.: Research project evaluation and selection: an evidential reasoning rule-based method for aggregating peer review information with reliabilities. Scientometrics, 105(3), pp. 1469–1490, 2015.

194. Zhu, J., Zhang, S., Chen, Y., Zhang, L.: A Hierarchical Clustering Approach Based on Three-Dimensional Gray Relational Analysis for Clustering a Large Group of Decision Makers with Double Information. Group Decision and Negotiation, 25, pp. 325–354, 2016.

195. Zulueta, Y., Martínez-Moreno, J., Bello, R., Martínez, L.: A discrete time variable index for supporting dynamic multi-criteria decision making, International Journal of Uncertainty, Fuzziness and Knowledge-Based Systems. 22, pp. 1–22, 2014.

Printed in the United States
By Bookmasters